Sustainable Energy Generation and Storage

Vijayanand Suryakant Moholkar ·
Kaustubha Mohanty · Vaibhav V. Goud
Editors

Sustainable Energy Generation and Storage

Proceedings of NERC 2022

Springer

Editors
Vijayanand Suryakant Moholkar
Department of Chemical Engineering
Indian Institute of Technology Guwahati
Guwahati, Assam, India

Kaustubha Mohanty
Department of Chemical Engineering
Indian Institute of Technology Guwahati
Guwahati, Assam, India

Vaibhav V. Goud
Department of Chemical Engineering
Indian Institute of Technology Guwahati
Guwahati, Assam, India

ISBN 978 981-99-2090-7 ISBN 978-981-99-2088-4 (eBook)
https://doi.org/10.1007/978-981-99-2088-4

This Springer imprint is published by the registered company Springer Nature Singapore Pte Ltd.
The registered company address is: 152 Beach Road, #21-01/04 Gateway East, Singapore 189721, Singapore

Foreword

It is a matter of great satisfaction for me that Indian Institute of Technology Guwahati successfully hosted North-East Research Conclave (NERC) 2022 during 20–22 May 2022. The NERC 2022 was conducted on the theme "Sustainable Science and Technology". Concurrently, Assam Biotech Conclave (ABC) was also organized on 21–22 May 2022. Both the events attracted huge participation from policy-makers, researchers, industrialist, army and students. Even the participation of school children was overwhelming.

NERC and ABC had many events including panel discussions, exhibitions, keynote lectures, competitions and paper presentations. Presentation of technical papers forms the core of any research conference. NERC attracted 879 research papers on various themes covering science, technology and humanities. Out of these, some select papers have been published by Springer Nature in the form of 15 volumes. These papers have been peer reviewed and thoroughly edited by IIT Guwahati faculty members. I am sure that these volumes will prove to be excellent resource material for research. Most of the papers presented in these volumes highlight the special needs and aspiration of eight states of North-East India. I congratulate and thank authors, reviewers, editors and publisher for bring out proceedings.

Motivation for organizing NERC came from none other than Honourable Minister of Education, Government of India, Shri Dharmendra Pradhan Ji. It helped to bring policy-makers, researchers, industrialists, academicians, students and children in one forum. It is perhaps the rarest conclave covering almost all possible research themes. For better readability, the proceedings has been divided into 15 volumes, but each volume reflects diversity in terms of topics and researchers. Only common thread is sustainable development of North-East India. Invariably, Sustainable North-East India is a prerequisite for sustainable India and the whole world. In that sense, these

15 volumes will serve guiding and stimulating light for all the stakeholders of the development. I am pleased to dedicate these volumes to nation as a part of Azadi ka Amrit Mahotsav.

T. G. Sitharam
Director
Indian Institute of Technology
Guwahati
Guwahati, India

Preface

As India progresses fast in the twenty-first century, we also face daunting challenges of energy security and climate change. Tremendous development in various sectors like industry, agriculture and transport has resulted in huge rise in demand for energy. Fulfilling these demands through conventional fossil fuel-based energy generation has given rise to significant emissions (both gaseous and liquids) that have caused pollution to atmosphere and aquatic eco-systems. Use of sustainable and green (or renewable) resources and technologies offers a viable and promising solutions to these issues. Sustainability can be defined as new-age societal goal with three dimensions: environmental, economic and social. Sustainable development means meeting the human needs without jeopardizing the integrity and stability of the natural systems and resources. One of the basic human needs is energy. The day-to-day energy needs of a common person are in three forms, viz. electricity, transportation fuel and domestic (cooking) fuel. The daunting issues of energy security and climate change risk due to GHG emissions have been the driving forces for the quest of renewable or sustainable energy resources. The issue of renewable or sustainable energy is more pressing for a developing country like India, who meets more than 80% of its oil needs through imports that creates tremendous burden on economy. On electricity front, although the installed generation capacity in India is nearly 400 GW (including the share of renewable energy of about 39%), the per capita electricity consumption is barely about 1200 kWh, which is one of the lowest in the world. It is high time for India to achieve self-reliance (*atmanirbharata*) in energy to become the leading global economy.

Last two decades have witnessed intense research activities in Indian academic institutions on renewable energy resources. These include biofuels (both liquid and gaseous) through thermochemical and biochemical conversion of biomass, solar energy through thermal and photo-voltaic routes, wind energy and hydroelectric energy. Government of India has supported R&D activities in renewable and sustainable energy through numerous schemes and projects. The provisions for renewable energy in the latest Union Budget 2022 include: (1) allocation of Rs. 19,500 crore for solar PLI scheme, (2) introduction of sovereign green bonds in public sector projects and (3) co-firing of 5–7% biomass pellets in thermal power plants that would save an estimated 38 MMT of CO_2 emissions annually.

The purpose of the session (track EG, No. 13) on Sustainable Energy Generation and Storage in the NERC 2022 was to bring together researchers in diverse fields of energy engineering and provide a knowledge-sharing platform. We received overwhelming response to call for papers. A total of 55 abstracts were received, and 49 of them were shortlisted for presentation. Despite adverse weather conditions during conclave, most of the participants travelled to IIT Guwahati to present their papers. The NERC itself turned out to be grand event with participation of more than 6000 delegates including school children. Three-day programme of NERC (20–22 May 2022) included technical sessions on numerous tracks of science and engineering, plenary and motivational talks by renowned personalities from industry and academia, a grand expo with participation from more than 100 start-ups, industries and academic/research institutions and ideation workshops for school children. The session on Sustainable Energy Generation and Storage, which had components of both oral and poster presentations, helped NERC participants to interact and exchange information and thoughts and stimulate new ideas. The interactions are expected to build networks and foster collaborations among academic and research institutions of North-East.

At this juncture, we would like to express our deepest gratitude to all faculty colleagues and students who made contributions to organization of this track of NERC. At the outset, we thank our Director, Prof. T. G. Sitharam and Dean R&D Prof. Vimal Katiyar for their patronship and pivotal role in organization of the mega event of NERC 2022. We also acknowledge Chairman of Technical Sessions Prof. U. S. Dixit for his coordination and guidance. We are thankful to all faculty members of School of Energy Science and Engineering, viz. Dr. Pankaj Kalita, Dr. Ranjith T., Dr. E. S. N. Raju and Dr. Lopakshi Barbora for their active participation in planning of session. We also express our gratitude to chairmen of different sessions, Prof. Senthilkumar, Prof. Ranjan Tamuli and Prof. N. Sahoo. We feel wordless to appreciate the enthusiasm and support of the student volunteers before, during and after the NERC 2022. We sincerely thank Karan Kumar, Rabindra K. Banik, Aparupa Thakuria, Nayanmoni Baishya, Avinash Anand, Kundan Kumar, Rohan Ghosh, Arunkumar M., Jaishree Bharadwaj and Saptaswa Biswas. Compiling of all manuscripts for the Springer volume of proceedings was also an important task. A total of 19 manuscripts were submitted initially, out of which 16 have been included in the final volume. We thank all reviewers of the manuscripts. Moreover, we also thank the student volunteers who helped in editing and formatting of the final draft of proceedings, viz. Karan Kumar, Avinash Anand and Harshendra N. Shet.

Finally, we hope that this Springer volume of proceedings on Sustainable Energy Generation and Storage will be a good source of information and knowledge to the researchers in this field.

Guwahati, Assam, India
November 2022

Vijayanand Suryakant Moholkar
Kaustubha Mohanty
Vaibhav V. Goud

About IIT Guwahati

Indian Institute of Technology (IIT) Guwahati established in 1994 has completed 25 years of glorious existence in 2019. At present, the Institute has eleven departments, seven interdisciplinary academic centres and five academic schools covering all the major engineering, science, health care, management and humanities disciplines, offering B.Tech., B.Des., M.A., M.Des., M.Tech., M.Sc. and Ph.D. programmes. The institute presently offers a residential campus to 435 faculty members and more than 7500 students at present. Besides its laurels in teaching and research, IIT Guwahati has been able to fulfil the aspirations of people of the North-East region to a great extent since its inception in 1994. The picturesque campus is on a sprawling 285 hectares plot on the north bank of the Brahmaputra, around 20 km from the heart of the Guwahati city.

IIT Guwahati is the only academic institution in India that occupied a place among the top 100 world universities—under 50 years of age—ranked by the London-based Times Higher Education (THE) in the year 2014 and continues to maintain its superior position even today in various International Rankings. IIT Guwahati gained rank 37 globally in the "Research Citations per Faculty" category and overall 384 rank in the QS World University Rankings 2023 released recently. IIT Guwahati has retained the 7th position among the best engineering institutions of the country in the "India Rankings 2021" declared by the National Institutional Ranking Framework (NIRF) of the Union Ministry of Education. IIT Guwahati has been also ranked 2nd in the "Swachhata Ranking" conducted by the Government of India. Recently, IIT Guwahati has been ranked as the top-ranked University in 2019 for IT developers by HackerRank in the Asia-Pacific region.

Among other frontier areas of research and innovation, IIT Guwahati is working towards augmenting critical science research initiatives in genomics, developmental biology, health care and bioinformatics, flexible electronics, advanced functional materials, sustainable polymers, rural technologies, renewable energy, artificial intelligence, disaster resilience and risk reduction and water resources and management. In its silver jubilee year, IIT Guwahati is poised to scale newer heights through all-round growth and development.

Indian Institute of Technology Guwahati has dedicated itself to the cause of improving and empowering Northeast India through cutting-edge research, region relevant projects, innovations, individual and multilateral collaborations and special initiatives. Being the only IIT in the entire Northeastern region, IIT Guwahati has an immense amount of responsibility to develop the region and empower the people of the region.

While the entire country is celebrating the "Azadi ka Amrit Mahotsav"—75 glorious years of Independence, and the great pride with which our nation of more than a billion people has been steadily growing today, IIT Guwahati is strongly committed to support that pace of growth for the entire NE so that we can keep pace along with the rest of the country. The specific areas of focus where IIT Guwahati has been contributing immensely to the region are:

(a) Infrastructure development across multiple sectors
(b) Providing solutions for multiple natural disasters such as recurring floods, landslides, earthquakes, cyclones, hailstorms and other natural calamities
(c) Improving the education sector and creating opportunities for employment
(d) Internet, telecommunication and cultural integration
(e) Technological intervention in interdisciplinary areas
(f) Healthcare services and education
(g) Renewable energy generation (solar, wind, biomass, hydro, geothermal)
(h) Overall industrialization, refining fossil fuels and setting up biorefineries.

Besides bringing in the state-of-the-art technical knowhow for most of the above sectors, the institute has been partnering with the local governments and enhancing the technological and educational interactions such that the next-generation youth are empowered with knowledge, skills and necessary entrepreneurial ability. These measures in Assam as well as all other northeast states will usher in a new era of growth, and the opportunities it will provide for interaction with the ASEAN countries as part of the Act East Policy of the Government of India will bring prosperity to this region.

Prof. Parameswar K. Iyer
Dean, Public Relations, Branding and Ranking
Indian Institute of Technology Guwahati

From the Desk of Chairman of Technical Committee of NERC 2022

North-East Research Conclave 2022 was successfully organized during 20–22 May 2022 with the participation of thousands of delegates. A total of 879 oral and poster papers were presented in the conference on 16 different tracks. The theme of the conclave was Sustainable Science and Technology, which is very pertinent in the modern era of globalization. Science and technology has to address economic, environmental and social problems of the world. Technology and sustainability are not incompatible. In fact, technology can achieve the goal of sustainability, which also includes preserving our rich cultural heritage. Concurrently with North-East Research Conclave (NERC), Assam Biotech Conclave 2022 was also organized on 21–22 May 2022. These mega events were organized at Indian Institute of Technology Guwahati (IITG) in physical mode after two years of pandemic period. Along with IITG, Science, Technology and Climate Change Department and Department of Education, Government of Assam were also organizers of these events under the patronage of Shri Dharmendra Pradhan Ji, Honourable Minister of Education and Minister of Skill Development and Entrepreneurship in the Government of India, and Shri Himanta Biswa Sarma Ji, Honourable Chief Minister of Assam.

It is a matter of great pleasure that Springer Nature is publishing the select papers from the conclave in 15 volumes. These are Advanced Functional Materials, Low Cost Manufacturing Technologies, Agro and Food Processing Technologies, Artificial Intelligence and Data Science based R&D interventions, Conservation of Biodiversity in the North Eastern States of India, Disaster Management, Healthcare Research and Related Technologies, Innovative Design for Societal Needs, Policies for Research and Innovation, Research and Innovation for Sustainable Development Goals, Sustainable Environment, Sustainable Energy Generation and Storage, Sustainable Transportation and Urban Development, Teaching and Learning Technologies, Technologies for Rural Development. These volumes are useful archival and reference materials for policy-makers, researchers and students.

As Chairman of Technical Committee, I am thankful to all Editors of all volumes, reviewers and student volunteers who have put tireless efforts to review, select and edit the papers of respective divisions, overcoming the time-constraint. Support provided by Convener, Prof. Vimal Katiyar, Dean R&D, IITG, and Co-conveners

Prof. Subhendu Sekhar Bag, Associate Dean R&D, IITG, and Shri Kailash Karthik N., IAS, is commendable. It is difficult to express words of gratitude for the Director, IITG, Prof. T. G. Sitharam, who has been motivating and guiding all the teams of NERC 2022 and ABC 2022.

Uday Shanker Dixit
Professor, Department of Mechanical Engineering, and
Head, Center for Indian Knowledge Systems

North East Research Conclave-2022: Toward Sustainable Science and Technology

It is extremely important and imperative to have knowledge-driven growth based on innovation in the case of academic higher education institutes of high repute. The North-Eastern region endowed with rich biodiversity comprises eight states. However, the climatic conditions, limited connectivity, lack of research infrastructure/institutes, territorial conflicts and the mountainous terrain of these regions are major impediments to the research ecosystem in the North-East. Quality higher education focusing on industry–academia collaboration and translational research is extremely beneficial for society. It has also been rightly pointed out by the Hon'ble Prime Minister Sh. Narendra Modi that, "*India cannot develop till Eastern India develops*".

With this idea and as India marks 75 years of Independence, Indian Institute of Technology Guwahati organized "The North-Eastern Research Conclave" from 20

to 22 May 2022. This grand event was jointly conducted with Science, Technology and Climate Change Department and the Department of Education, Government of Assam, at IIT Guwahati Campus.

The mission behind the conclave was to showcase the best R&D activities from educational and research institutions across North-East India and to create an environment, conducive to development of local indigenous technologies and innovations, creating the scope and laying the foundation for entrepreneurship.

In order to attract people and spread awareness about the event, a roadshow was initiated from IIT Guwahati on 7 May 2022 in order to reach all the partnering academic institutes and make them an integral part of the mega event. The Director, IITG, waved the NERC 2022 flag and sent off the road show vehicle from the institute. More than 400 students, staff and faculty participated actively in the roadshow.

A huge response was received by participants from throughout the country. The total no. of participating institutions in this conclave included 7 IITs, 10 NITs, 5 IIITs and other CFTIs, 23 research laboratories, 17 central-funded universities, 47 other universities/institutes along with about 100 schools. Eminent personalities from industries, start-ups, research councils and PSUs also joined in.

The presence of dignitaries from important Ministries was observed such as Shri Dharmendra Pradhan, Hon'ble Union Minister of Education and Minister of Skill Development and Entrepreneurship, Government of India; Dr. Himanta Biswa Sarma, Hon'ble Chief Minister of Assam State; Dr. Ranoj Pegu, Hon'ble Minister of Education, Government of Assam; Dr. Rajkumar Ranjan Singh, Hon'ble Minister of State for Education, Government of India; Dr. Subhas Sarkar, Hon'ble Minister of State for Education, Government of India; Shri Keshab Mahanta, Hon'ble Minister of Science Technology and Climate Change, Government of Assam and many more.

The inauguration ceremony of the conclave was followed by the signing of an MoU between IIT Guwahati and the Government of Assam to establish "The Assam Advanced Health Innovation Institute (AAHII)". This MoU would prove to be a unique partnership between the Government of Assam and IIT Guwahati in order to set up a research institution to leverage advanced technologies to transform medical science. This joint venture company will be able to invite participation from intending parties including corporates/businesses/research institutions and philanthropic organizations.

The third edition of Assam Biotech Conclave 2022 was also held as part of NERC 2022. It brought together the Biotech Entrepreneurs, industry leaders, researchers, academicians, government representatives, policy-makers, innovators and investors together on one platform to explore the possibilities of biotechnology in North-East India and to discuss the new opportunities in the transition.

Officers from the Indian Army also actively participated in the conclave. A talk on "Atmanirbhar Bharat—Indian Army Initiatives towards Self Reliance" was delivered by Lt. Gen. D. S. Rana AVSM, YSM, SM General Officer Commanding, Gajraj Corps on 21 May 2022. The talk was aligned with the vision of the apex leadership of the Government of India and initiatives undertaken by the Indian Armed Forces with a focus on the integration of civil–military establishment in the field of self-reliance. He also elucidated that institutions such as IIT Guwahati which has many running research projects and elaborate student exchange and joint collaboration setup with a large number of countries have the wherewithal to take up defence-related R&D and also facilitate delivery with industry partners. He also invited IIT Guwahati to participate in EAST TECH Symposium planned at Kolkata in July 2022. This led to the signing of an MoU between Indian Army Eastern Command and IIT Guwahati on 7 July 2022 during East Tech 2022. This would further impetus to Indigenisation and Raksha Atmanirbharta.

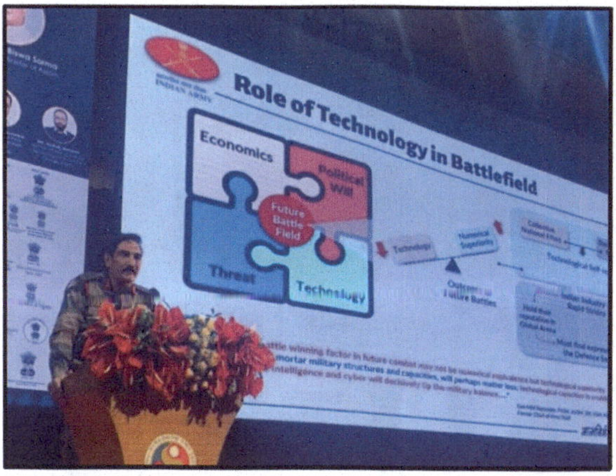

Global battery experiment was performed by more than 1300 students in three sessions starting from 20 May to 22 May at IIT Guwahati. Along with the global battery experiment, creating skilful educators (teacher training programme) was also conducted in parallel sessions. Students had arrived from various schools across Assam and other North-Eastern states.

The Guwahati Declaration was launched at the valedictory ceremony of the conclave by Shri Lok Ranjan, Secretary, Ministry of Development of North Eastern Region (DoNER), in the presence of Shri Kailash Karthik, Deputy Commissioner, Kamrup. The declaration is intended to create a set of guidelines, through which individual as well as a collective responsibility to promote and encourage innovation at the grass-root level and strive to stimulate and execute indigenization and entrepreneurship can be taken up.

Science, education, research and innovation are the four pillars on which the development, as well as the work culture of a nation, rests. This was well articulated by the promising number of exhibitors being seen participating from all across the NE states in the NERC 2022. All the NITs, CFTIs and CFIs were allocated two stalls each, where the delegates showcased the working models of their inventions. Distinctive pavilions were arranged for IIT, NIT, CFIs and CFTIs. Excellent response was obtained from the start-ups all across the NE states. Federation of Industry Commerce of North Eastern Region (FINER) had partnered with NERC 2022 as an industry partner, and they showcased 50 start-ups as a part of the exhibition under the FINER Pavilion. Other significant organizations that came forward to showcase their allied R&D start-ups were the Oil and Natural Gas (Oil and Natural Gas Pavilion), Indian Army (Defense Pavilion) and NE-Railway (NE-Railway Pavilion).

Multifarious research work on topics of societal relevance was presented by researchers from different organizations/institutes. The presentations were conducted in oral and poster presentation modes. The thematic areas for these presentations were part of some of the Sustainable Development Goals (SDGs) such as SDG-3: Good Health and Wellbeing; SDG-7: Affordable and Clean Energy; SDG-9: Industry, Innovation and Infrastructure; SDG-11: Sustainable Cities and Communities and SDG-12: Responsible Consumption and Production. Some of the papers highlighted environmental sustainability, efficiency and management issues, which are important to be presented in the case of North-East regions. Two awards were given under each technical category for these presentations. Overall, the technical sessions were a grand success due to the active cooperation from editors, chairpersons of all the sessions and student volunteers of IITG.

The Government of India has taken various steps to encourage women in the field of science and technology. In this line, the IIT Guwahati Woman Researcher Award was approved to recognize the contribution of women Faculty members of IIT Guwahati fraternity. This prestigious award was conferred to Dr. Latha Rangan who is Senior Professor in the Department of Biosciences and Bioengineering, Indian Institute of Technology Guwahati, India. Prof. Rangan has played a key role in plant biotechnology and sustainable development and especially in the areas of energy security, food security and medicinal crops.

The conclave paved the way for creating mass awareness of Research and Innovation for developing a sustainable society. There was knowledge exchange and dissemination that led to the establishment of Centres of Excellence in Translational Collaborative Research and Innovation. This mega event led to the bridging of the gap between industry–academia and creating handholding pathways for setting up long-term collaboration for R&D innovations towards the goal of establishing sustainable NE India. The conclave brought together over 8000 participants including Hon'ble Ministers, Official Bureaucrats, Eminent Professors, Scientists, Renowned Industrialist, School Children/Teachers and Others delegates. This revolutionized the R&D road map of all the NE states through various dissemination of policies which will benefit the sustainable development of all NE states in near future.

It is an honour and a moment of extreme pride for getting the NERC proceedings published in the prestigious Springer volumes. We would like to thank and acknowledge the globally active publisher Springer for helping us being able to publish the articles on 15 broad areas. We would also like to thank all the authors for their contribution to the grand success of NERC 2022 and wish them great success in all of their future endeavours.

Prof. Vimal Katiyar
Dean, R&D
Department of Chemical Engineering
Centre for the Sustainable polymer
Indian Institute of Technology Guwahati
Guwahati, India
vkatiyar@iitg.ac.in

Prof. Subhendu Sekhar Bag
Associate Dean, R&D
Department of Chemistry
Centre for the Environment
Indian Institute of Technology Guwahati
Guwahati, India
ssbag75@iitg.ac.in

Contents

Editors and Contributors

About the Editors

Prof. Vijayanand Suryakant Moholkar (b. 1972) is a Professor (HAG) of Chemical Engineering and Adjunct Faculty at School of Energy Science and Engineering (erstwhile Centre for Energy) at Indian Institute of Technology (I.I.T.) Guwahati. He received B.Chem.Engg. (1993) and M.Chem.Engg. (1996) degrees in chemical engineering from Institute of Chemical Technology (formerly University Department of Chemical Technology, UDCT) Mumbai, followed by Ph.D. from University of Twente (Netherlands) in 2002. He has been Head of the Chemical Engineering Department, IIT Guwahati between 2012–2015, and Head of Centre for Energy, IIT Guwahati between 2017–2020. His main research interests are sonochemistry, cavitation assisted physical, chemical and biological processing, and thermo- and biochemical routes to biofuels. As of January 2023, he has published more than 195 papers in renowned international journals that have received more than 10,500 citations (with h-index of 60 and i-10 index of 170). He is co-inventor of 3 US patents (issued to CTI Nanotech, CA, USA) on application of hydrodynamic cavitation reactors for biomass pretreatment and bioalcohol synthesis. As of January 2023, he has graduated 26 Ph.D. and 34 M.Tech. students. He has been elected as Fellow of Royal Society of Chemistry (FRSC) in July 2016. He has also been elected as Fellow of Institution of Chemical Engineers, UK. He is a Chartered Engineer (C.Eng.) registered with Engineering Council of UK. He was admitted as Senior Member of American Institute of Chemical Engineers (MAIChE (Sr.)) in August 2016. His name has consistently featured in the list of world's top 2% scientists published by Stanford University, USA for the years 2020, 2021 and 2022.

Dr. Kaustubha Mohanty is currently working as Professor and Head in the Department of Chemical Engineering and as Adjunct Professor in the School of Energy Science and Engineering at IIT Guwahati. He completed his Ph.D. from IIT Kharagpur and worked as Post-Doctoral researcher in Ben-Gurion University, Israel and McMaster University, Canada. He has published 185 research articles in international peer-reviewed journals that have received more than 7500 citations (with an h-index of 47). Till date, Prof. Mohanty has guided 20 Ph.D. students at IIT Guwahati. Presently, he is supervising another 15 Ph.D. students in diverse fields of Biofuels, Microalgae Biorefinery, Biomass Pyrolysis, Biological Wastewater Treatment, Membrane Technology and Heterogeneous Catalysis. He has completed several research and consultancy-based projects sponsored by several funding agencies such as MHRD, DRDO, DST, CSIR, ONGC and DBT. He is an active member of several professional organizations such as Fellow of the Royal Society of Chemistry, London; Fellow of Biotech Research Society, India; Fellow of Institution of Engineers, India; Member of Society of Chemical Industry, London; Member of Chemical Institute, Canada and Life Member of Eco-Ethics International Union, Germany.

Dr. Vaibhav V. Goud is currently working as Professor and Head in the School of Energy Science and Engineering at IIT Guwahati. He completed his Ph.D. from IIT Kharagpur in 2006. He worked as Lecturer at BITS Pilani, Goa Campus between 2007–2008, and between 2008–2009 worked as a postdoctoral fellow at the University of Saskatchewan, Saskatoon, Canada. Till date, Prof. Goud guided 14 Ph.D. students at IIT Guwahati. Presently, he supervising 09 Ph.D. students in diverse fields of biomass for the production of biofuels, bio-lubricants and oleo-chemicals, extraction of natural plant products (essential oils, oleoresins, food colors, biopesticides, and nutraceuticals) by using subcritical and supercritical fluids as well as the processing of non-edible oilseeds. He has published more than 150 papers in international peer-reviewed journals that have received more than 10,000 citations and 18 book chapters and made presentations of his research at several national/international conferences. He has been awarded/filed 3 Indian patents in the field of bioethanol, biodiesel, and enhanced recovery of lipid from microalgae. He is an active member of several professional organizations such as life member of the Indian Institute of Chemical Engineers, Sea buckthorn Association of India and International Society of Food Engineering.

Contributors

Pratima Agarwal Department of Physics, Indian Institute of Technology Guwahati, Guwahati, Assam, India

Gaffer Ahmed Department of Chemical Engineering, Indian Institute of Technology Guwahati, Guwahati, Assam, India

Avinash Anand Department of Chemical Engineering, Indian Institute of Technology Guwahati, Guwahati, Assam, India

Praveen Kumar Reddy Annapureddy Department of Chemical Engineering, Indian Institute of Technology Guwahati, Guwahati, Assam, India

Punampriya Borgohain Department of Chemical Engineering, Indian Institute of Technology Guwahati, Guwahati, Assam, India

Himangshu Deka Department of Electronics and Instrumentation Engineering, National Institute of Technology Silchar, Silchar, Assam, India

Raghvendra Gupta Department of Chemical Engineering, Indian Institute of Technology Guwahati, Guwahati, Assam, India;
Center for Sustainable Polymers, Indian Institute of Technology Guwahati, Guwahati, Assam, India

Mohammad Anzar Hussain Department of Chemical Engineering, Indian Institute of Technology Guwahati, Guwahati, Assam, India

Pankaj Kalita School of Energy Science and Engineering, Indian Institute of Technology Guwahati, Guwahati, Assam, India

Nanda Kishore Department of Chemical Engineering, Indian Institute of Technology Guwahati, Guwahati, Assam, India

Vinayak Kulkarni Department of Mechanical Engineering, Indian Institute of Technology Guwahati (IITG), Guwahati, India

Karan Kumar School of Energy Science and Engineering, Indian Institute of Technology Guwahati, Guwahati, Assam, India

Juhi Kumari School of Energy Science and Engineering, Indian Institute of Technology Guwahati, Guwahati, Assam, India

Deepak Kumar Mishra Department of Chemical Engineering, Indian Institute of Technology Guwahati, Guwahati, Assam, India

Vijayanand S. Moholkar Department of Chemical Engineering, Indian Institute of Technology Guwahati, Guwahati, Assam, India;
School of Energy Science and Engineering, Indian Institute of Technology Guwahati, Guwahati, Assam, India

P. Muthukumar School of Energy Science and Engineering, Indian Institute of Technology Guwahati, Guwahati, Assam, India;
Department of Mechanical Engineering, Indian Institute of Technology Guwahati, Guwahati, Assam, India

Puneet Kumar Nema School of Energy Science and Engineering, Indian Institute of Technology Guwahati, Guwahati, Assam, India

Rahul Department of Physics, Indian Institute of Technology Guwahati, Guwahati, Assam, India

Draksharapu Rammohan Department of Chemical Engineering, Indian Institute of Technology Guwahati, Guwahati, Assam, India

Umang H. Rathod Department of Mechanical Engineering, Indian Institute of Technology Guwahati (IITG), Guwahati, India

Ujjwal K. Saha Department of Mechanical Engineering, Indian Institute of Technology Guwahati (IITG), Guwahati, India

Harshendra N. Shet K School of Energy Science and Engineering, Indian Institute of Technology, Guwahati, India;
Department of Electrical and Electronics Engineering, Manipal Institute of Technology, Manipal Academy of Higher Education, Manipal, India

Anugrah Singh Department of Chemical Engineering, Indian Institute of Technology Guwahati, Guwahati, Assam, India

Gaurav Singh School of Energy Science and Engineering, Indian Institute of Technology Guwahati, Guwahati, Assam, India

Arun Kumar Sunaniya Department of Electronics and Instrumentation Engineering, National Institute of Technology Silchar, Silchar, Assam, India

Ranjith Thangavel School of Energy Science and Engineering, Indian Institute of Technology Guwahati, Guwahati, Assam, India

Narra Thejaswini Department of Chemical Engineering, Indian Institute of Technology Guwahati, Guwahati, Assam, India

Pankaj Tiwari Department of Chemical Engineering, Indian Institute of Technology Guwahati, Guwahati, Assam, India

Rajesh Kumar Upadhyay Department of Chemical Engineering and Technology, Indian Institute of Technology (BHU) Varanasi, Varanasi, Uttar Pradesh, India

R. V. S. Uppaluri Department of Chemical Engineering, Indian Institute of Technology Guwahati, Guwahati, Assam, India

Vijaya School of Energy Science and Engineering, Indian Institute of Technology Guwahati, Guwahati, Assam, India

Development of a Novel Drag-Based Vertical-Axis Wind Rotor Inspired from Orange Sea-Pen

Umang H. Rathod, Vinayak Kulkarni, and Ujjwal K. Saha

Abstract For the purpose of harvesting wind energy from low wind velocity regions such as north-east India, the Savonius-type drag-based vertical-axis wind turbine (VAWT) can be a potential candidate due to direction independency, absence of yaw mechanism, small-scale stand-alone systems and ease of fabrication due to absence of airfoil-shaped blades. Based on the aforementioned merits, it is decided to conduct a performance improvement task by developing a novel blade shape bio-inspired from the polyp leaf of the orange sea-pen (*Ptilosarcus gurneyi*). The similarities between flow physics of the Savonius rotor and the feeding mechanism of the sea-pen are discussed. The procedure of shape extraction for novel blade is narrated and the rotors are fabricated. Wind tunnel experimentation and 2D numerical simulations are performed for preliminary performance assessment of the rotors. The performance potential of the novel blades is assessed in comparison with the semicircular blade. The result indicated the possibility of the performance improvement in case of novel bio-inspired blade.

Keywords Vertical-axis wind turbine Savonius wind rotor · Bio-mimicry · Orange sea-pen

1 Introduction

In order to harvest wind energy, two types of wind turbines are generally employed, namely horizontal-axis wind turbine (HAWT) and vertical-axis wind turbine (VAWT). HAWTs provide higher efficiency in comparison to VAWTs; however, the latter ones have become more popular due to their unique characteristics such as ability of operating in low wind speed, direction independency, absence of yaw mechanism and stand-alone small-scale energy generating system. In this category, drag-based Savonius-type wind rotor is among one of the most prevalent VAWTs

U. H. Rathod (✉) · V. Kulkarni · U. K. Saha
Department of Mechanical Engineering, Indian Institute of Technology Guwahati (IITG), Guwahati, India
e-mail: umang174351003@iitg.ac.in

V. S. Moholkar et al. (eds.), *Sustainable Energy Generation and Storage*,
https://doi.org/10.1007/978-981-99-2088-4_1

such as Darrius rotors, Gorlov rotors or Φ—rotor, due to the absence of airfoil-shaped blades and hence ease in fabrication [1]. As the Savonius rotor requires low starting torque and can be operated in low wind speed area, it can be beneficial for north-east region of India and other parts of the world. The same design of the Savonius wind rotor has also been implemented as hydrokinetic rotor, successfully. As shown in Fig. 1, blades of a typical Savonius rotor are made of two halves of a cylinder. Among them, the blade with its concavity facing upstream of the wind flow is named as 'advancing blade', and the remaining one is 'returning blade'. Due to difference in the inherited shapes of both the blades, i.e., concavity and convexity, the drag force experience by the advancing blade is higher than the returning blade resulting into net positive torque as per the sign convention depicted in Fig. 1.

In order to improve the performance of the rotor, researchers have proposed different augmentation techniques which attempt to improve and diminish the drag forces experienced by the advancing and returning blades, respectively. Techniques such as guide box, deflector plates and wind shields block the upstream flow of the wind on the returning blade causing the drag force experienced by the returning

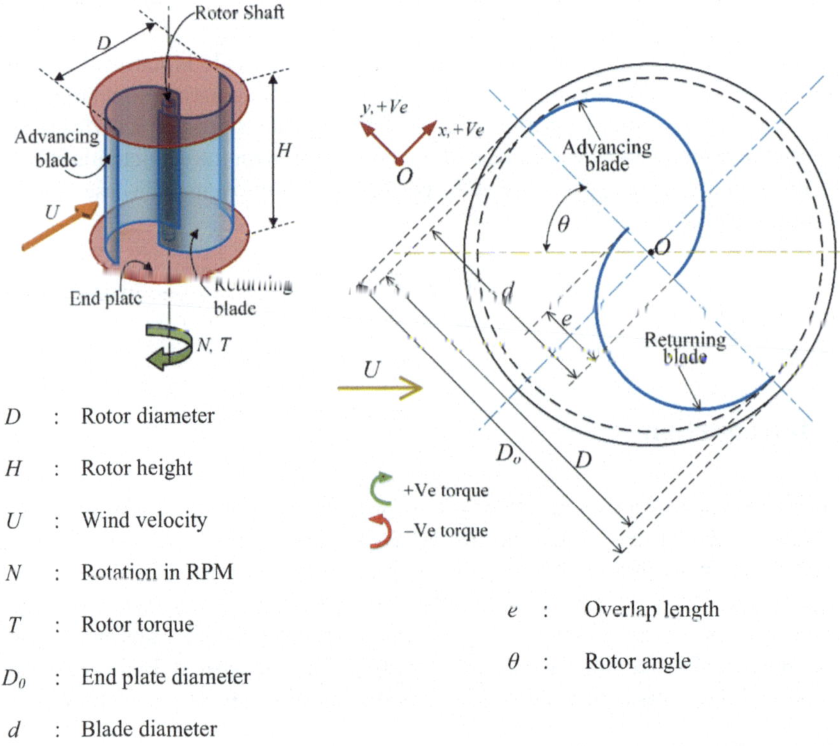

(a) Illustration of rotor parts **(b)** Geometric details

Fig. 1 A typical two-bladed Savonius rotor

blade [2]. As an extension to this technique, it has been proposed to block the flow at upstream of the returning blade and divert it toward the advancing blade, simultaneously. The examples of such augmentation techniques are concentrators, curtain plates, guiding path in form of ducts, nozzle duct and others [3]. The disadvantage of these techniques is that they eliminate the characteristics of the direction independency of the rotor, even though they improve the rotor performance. Therefore, instead of diverting the wind flow from the returning blade, alternative augmentation techniques were proposed which leak the stagnant flow from convex side (upstream) to the concave side (downstream) of the returning blade, e.g., slatted blades, hinged flaps, vented and capped vented blades, slotted blades and valve-aided blades [4]. However, it is reported that these techniques fail to improve the rotor performance. There are techniques, viz. number of blades and stages, blade twist, end plates and blade shapes, which do not affect both the direction independency and the rotor performance. Out of them, finding the optimum blade shape performing better than the conventional semicircular blade still remains the research interest as well as current trend. Researchers have opted for optimization techniques [5, 6], existing low drag aerodynamic shapes [7] and bio-mimicry-based techniques [8] to synthesize optimum blade shape. This trend motivated the authors to study the primary flow physics of the rotating Savonius rotor, gaining the insight and then search for appropriate bio-inspired phenomenon that can lead to improved blade shape.

Subsequently, the flow physics of the Savonius rotor is discussed along with analogous bio-inspired phenomenon in Section 1.2, followed by research methodologies in Section 1.3 describing experimental and numerical setups along with design of novel blade shape. Thereafter, primary experimental and numerical results are presented in Section 1.4. Lastly, the conclusions and future scope are listed in Section 1.5.

2 Flow Physics of the Savonius Rotor and Analogous Phenomena

In literature, the reported studies based on the flow physics are rather scanty as compared to numerical and experimental performance investigations. It is important to note that the drag forces (responsible for rotor torque) experienced by both the blades mentioned in the Section 1.1 are the result of the pressure difference between their concave and convex sides. In this context, Fujisawa et al. experimentally investigated the pressure distribution over all the surfaces of the blades at the rotor mid-height [9–11]. It was shown that the low-pressure region named as 'Coǎnda flow region' located toward the tip of the downstream advancing blade is very important for the torque generation in the rotor (Fig. 2). This low-pressure region is then formed into the clockwise rotating vortex and then detaches at higher rotor angles (θ). The remaining important flow regions are, namely the pressure recovery region at downstream side of the advancing blade and two stagnation regions at upstream sides of the blades as shown in Fig. 2 for $\theta = 90°$. Further, Rathod et. al. studied the

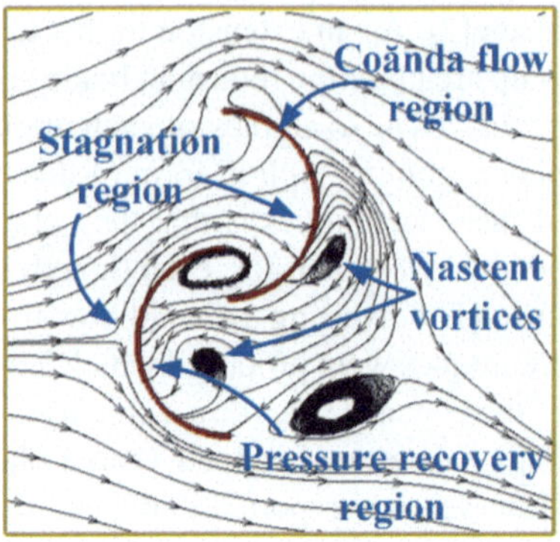

Fig. 2 Illustration of flow field during rotation at $\theta = 90°$

distribution of pressure and instantaneous local torque produced by each infinitesimal blade elemental lengths numerically. It was demonstrated that the low-pressure region toward the convex side of advancing blade tip contributes the most in terms of generating instantaneous local torque and overall torque of the rotor as compared to the other parts of the blades for majority of the θ values [4]. Therefore, it can be summarized that the focus should be on the shape of the advancing blade's tip, low-pressure region and vortex attached to the same in order to improve the rotor performance.

Based on the insight gained in the last paragraph, the authors of the present paper explored different biological phenomena, which take the advantage from pressure difference, drag and attached vortices. It is found that the bio-organism such as 'sea-pen' (order: *Pennatulacea*) (Fig. 3) uses pressure drop and attached vortices over their bodies for the purpose of feeding [12]. Sea-pen is the member of the octocorals, which remains mounted on the sea-floor and forages on the planktonic organisms without actively moving its major body parts. However, its tentacle like 'polyps' located on leaf-shaped appendages can move and capture planktonic food particles, making them passive feeders. This organism is mainly found in habitat that is under influence of uni-directional sea-currents and orients its concavity toward the upstream flow [13]. In order to maximize its feeding capacity, it maximizes its projected area by expanding its concave body, resulting into high pressure drop across the upstream and downstream side of the body and hence increase in the drag force. This pressure drop facilitates the planktonic food particles to flow through the polyp leaves from upstream to the downstream of the sea-pen body, which are then captured by the polyps. However, the majority of the polyps are located on the downstream side of the sea-pen body instead of upstream side, even though it looks like that the successful entrapment of food particles by polyps is most probable at upstream side rather than the downstream side. The reason is that the polyps enhance their particle capturing mechanism with the aid of vortices induced at downstream of sea-pen body. Moreover, it is also found that the concave shape of the polyp leaf enhances the strength of these attached vortices at downstream, which trap the nearby food particles into the downstream circulation and then catch them with the help of

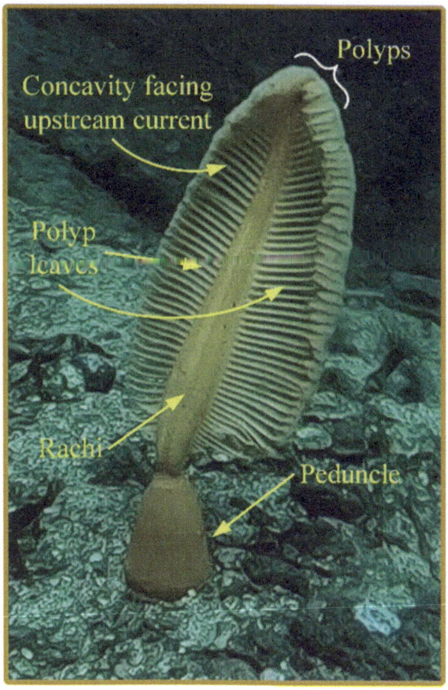

Fig. 3 Morphology of a sea-pen (orange sea-pen—*Ptilosarcus gurneyi*) [16]

polyps located downstream [14, 15]. Thus, the concave body shape as a result of the long evolutionary process may be optimum for producing high pressure drop, tip vortices and drag to increase feeding efficiency of the sea-pen. This matches the requirement of the proposed optimum blade shape of the Savonius rotor discussed in the first paragraph of this section.

The aforementioned similarity between flow physics of the Savonius rotor and the feeding mechanism of sea-pen suggests to extract the shape of the polyp leaf for a novel blade shape. It is important to note that the working environment for sea-pen is static and not a rotating one like Savonius rotor. However, it can be argued that the working mechanism of the Savonius rotor was deduced by considering the static drag difference between concave and convex shaped cups of conventional anemometer arms [17]. The same logic can also be extended to support the present analogy.

3 Research Methodology

This section contains discussion on blade shape extraction (Sect. 1.3.1), numerical and experimental setups (Sects. 1.3.2 and 1.3.3) and definition of performance parameters to evaluate the rotor performance (Sect. 1.3.4).

3.1 Blade Shape Extraction and Rotor Design

There are many sea-pen species having various sizes and shapes found in the nature. Among them one of the largest and well-studied species is the orange sea-pen (*Ptilosarcus gurneyi*), which can be found in the flow conditions of *Re (Reynolds number)* = 10^5 to 10^6 [18]. This range of *Re* is identical to the Savonius rotor with approximate size of $D \approx 250$ mm (and $H \approx 250$ mm) under operating condition of U = 4–6 m/s, which is the approximate wind condition for areas with low wind velocities. Moreover, size of the adult specimens of orange sea-pen is found in-between 400 and 500 mm in height with the diameter approximately half of its height. Due to this similarity, it is decided to extract the shape from the fully grown polyp of the orange sea-pen. Both the upstream and downstream shapes of the polyp leaf on planer view (top view of entire sea-pen) should be extracted. However, the former one collects food particle by maximizing the projection area, resulting into high pressure drop and drag force as discussed in Sect. 1.2. Similarly, the downstream shape is responsible for the enhanced attached vortices. However, the latter one is more important since it addresses the attached vortices on the tip. Further, the sea-pen detaches itself from ground and drifts with the sea-current, whenever it feels disturbed and frightened by the predators as shown in Fig. 4c [19]. The aim of the drifting sea-pen can be the maximization of drag force acting on its concave side because of sea-current. Thus, the upstream leaf shape is important not only for generating pressure drop but also for displacement of the same along with the sea-currents.

As the sea-pen tends to deflate or be carried away in the direction of sea-currents when being disturbed, physical measurement of the shape of its polyp leaf is not possible. Therefore, it is necessary to extract the shape digitally from the research

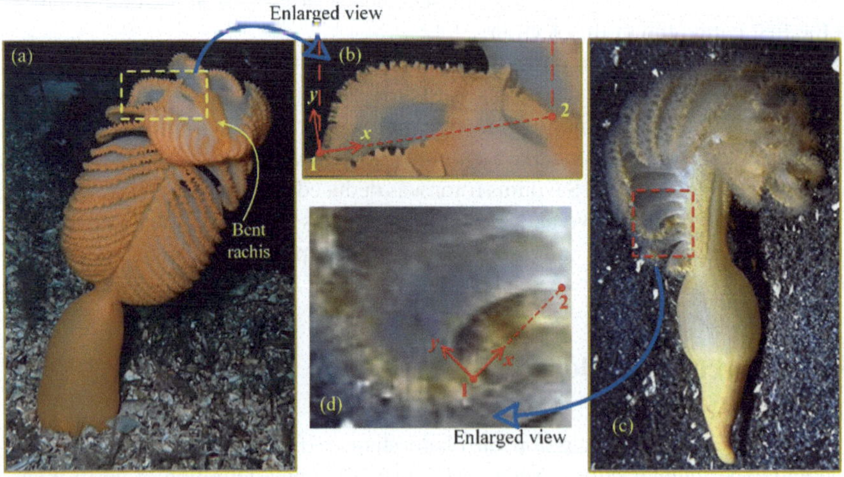

Fig. 4 **a**, **b** Shape extraction for profile 2 [20]. **c**, **d** Shape extraction for profile 1 [19]

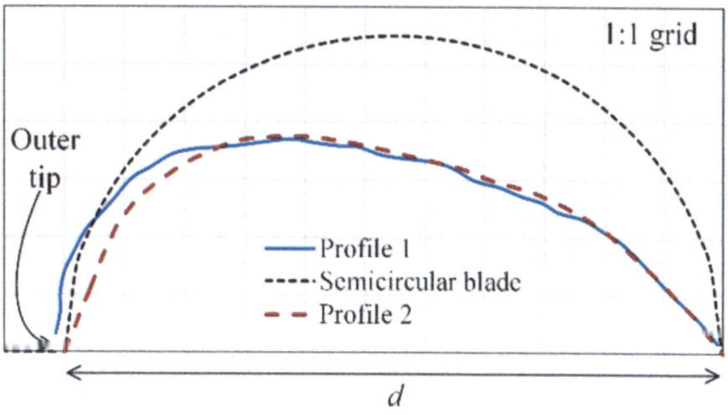

Fig. 5 Comparison among semicircular blade, profiles 1 and 2

grade images. Such images are showed in Fig. 4a, b and Fig. 4c, d, for downstream (profile 2) and upstream (profile 1) shapes, respectively. In order to extract the shape, image processing programming of *MATLAB 2021a* is implemented [21]. The procedure includes image cropping of individual polyp leaf, followed by grayscale conversion and lastly, the edge detection algorithm. Thereafter, the images are digitized by setting coordinates systems as per Fig. 4 and setting the chord of the polyp shape equal to the that of rotor blade (with end-points 1 and 2). Both the shapes along with conventional semicircular blade are plotted in Fig. 5 for comparison.

After extracting the blade shape, three rotors consisting of the aforementioned blade shapes are designed with $D \approx H \approx 250$ mm, without considering rotor shaft for numerical analysis. For experimental analysis, rotors having blades with profiles 1, profile 2 and semicircular shape are fabricated. As profile 2 has the influence over the attached tip vortices of sea-pen body, it is decided to give more importance to the same for the experimental investigations. Other dimensions are similar to that of numerical setup except the provision of the rotor shaft.

3.2 *Numerical Setup*

Due to consistency of the rotor cross section throughout the rotor height, it is decided to carry out 2D simulation instead of 3D simulation using *ANSYS FLUENT* software [22] since the latter one is computationally more exhaustive. The overall computational domain is divided into two parts as shown in Fig. 6, viz. rotating circular domain and stationary rectangular domain. The size of the domains and boundary conditions are specified in the Fig. 6. This domain size is finalized by conducting domain independence test [23–25]. Similar domain size was successfully exercised in documented studies [4]. In order to capture boundary layer on the rotor blade surfaces, inflation layers with sufficient thickness are applied by considering $y^+ <$ 1 (non-dimensional distance). 2D pressure-based solver along with *k*-shear-stress transport (SST) model [26–28] and semi-implicit method for pressure-linked equations (SIMPLE) algorithm is used to simulate the present transient phenomenon. Second-order scheme is used to solve both the temporal and spatial terms of the

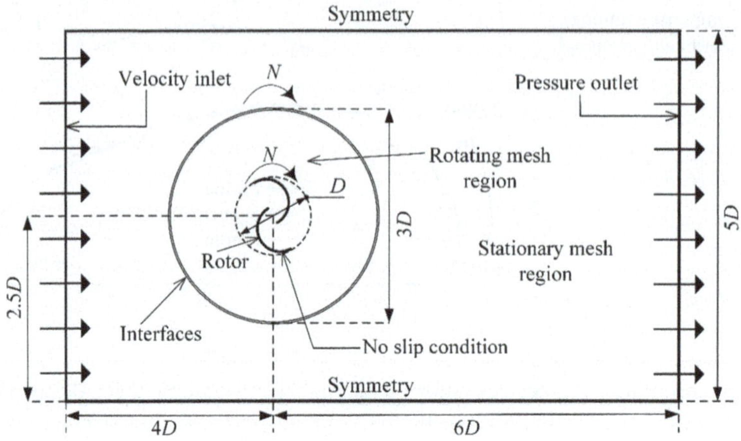

Fig. 6 2D computational domain and boundary conditions

governing equations. By performing grid independence test, the mesh structure having overall mesh elements approximately equal to 1.5×10^5 with average element quality of 0.94 is selected for the simulations. Further, the mesh structure possesses an average skewness ≈ 0.08 and a maximum cell aspect ratio ≈ 26, which are well below the maximum limits recommended as per the *ANSYS FLUENT* guidelines to ensure ease of numerical convergence [29].

3.3 Experimental Setup

The primary performance investigation is carried out in open-circuit subsonic wind tunnel facility at IIT Guwahati. The dimensions of the closed cross section of the wind tunnel is 600 mm $\times$ 600 mm $\times$ 2000 mm. The rotor along with the shaft is mounted in the test section with the help of bearing fixed on the top and bottom surface of the test section. The rope-brake type dynamometer is attached with the shaft extension to measure the torque produced by the turbine. The proximity sensor is used to measure the rotational speed of the rotor. The wind velocity (U) is measured using the electronic manometer with a pitot tube.

3.4 Performance Parameters

The torque (T) is measured using the dynamometer, and rotor's rotation per minute (N) is measured by the proximity sensor. Using these values, tip-speed ratio (λ), torque coefficient (C_T) and power coefficients (C_P) are calculated using the Eqs. 1 through 3, respectively. It is a convention that the performance of two rotors is compared in the form of C_P and λ.

$$\text{Tip - speed ratio, } \lambda = \frac{\omega D}{2U} = \frac{\pi D N}{60U} \quad (1)$$

$$\text{Torque coefficient, } C_T = \frac{T}{\frac{1}{2}\rho A U^2 \times \frac{D}{2}} \quad (2)$$

$$\text{Power coefficient, } C_P = C_T \times \lambda = \frac{T\omega}{\frac{1}{2}\rho A U^3} \quad (3)$$

4 Results and Discussion

The preliminary experimental and numerical investigation was carried out to assess the performance of the newly developed sea-pen blades as compared to conventional semicircular blade. As mentioned previously, numerical simulations for semicircular blade, profile 1 and profile 2 are carried out at $U = 7$ m/s or $Re = 1.2 \times 10^5$. On the other hand, preliminary experimental analysis is carried out for semicircular blade and profile 2 at $U = 7$ m/s ($Re = 1.2 \times 10^5$). If promising results are obtained in the present preliminary investigation, then the detailed performance analysis of the sea-pen bladed rotor should be carried out in the future.

4.1 *Numerical Investigation*

For this purpose, numerical simulation for three conditions, viz. $\lambda < 0.5$, $\lambda \approx 0.5$ and $\lambda > 0.5$, are carried out, and their values are provided in Table 1. From the table, it can be observed that the performance of both the profiles 1 and 2 is higher than that of the semicircular blade approximately at $\lambda < 0.5$ and $\lambda \approx 0.5$. However, the trend is reversed for $\lambda > 0.5$. Similar type of performance trend was also observed for Bach-type blade in comparison with the semicircular blade [30]. It is important to clarify that no significant difference in the overall performance was observed between profiles 1 and 2. Thus, the preliminary findings at $U = 7$ m/s show that the new blade profiles have potential to perform better than the semicircular blade.

4.2 *Experimental Investigation*

Experiments at three conditions, viz. $\lambda < 0.5$, $\lambda \approx 0.5$ and $\lambda > 0.5$, are carried out, and their values are provided in Table 2. It is quite evident that the experimental results

Table 1 Preliminary numerical dataset

λ	C_P		
	Semicircular blade	Profile 1	Profile 2
0.26	0.11	0.15	0.15
0.35	0.15	0.22	0.21
0.62	0.25	0.28	0.28
0.88	0.29	0.27	0.28

Table 2 Primary experimental dataset

λ	C_P	
	Semicircular blade	Profile 2
0.18	0.02	0.05
0.38	0.05	0.07
0.53	0.07	0.07
0.70	0.06	0.05

are supporting the numerical findings and the profile 2 performs better than the conventional semicircular blade. It is important to clarify that the numerical values of C_P are highly overestimated as compared to the experimental results. The reason behind this is the 2D simulation with k–ω SST model usually overestimates. Similar kind of overestimation is also observed in other reported wind turbine studies [24, 26].

5 Conclusion

From the preliminary experimental and numerical analysis of the semicircular blade, profile 1 and profile 2, following conclusions can be drawn out:

(a) Both the profiles 1 and 2 show possibility of better performance as compared to semicircular blade at $\lambda < 0.5$ and vice versa for $\lambda > 0.5$.
(b) No substantial performance difference is observed between profiles 1 and 2.
(c) The analogy established between the Savonius rotor and the sea-pen polyp is logical and is also supported by the preliminary experimental and numerical findings.

Since the performance improvement is witnessed by bio-mimicry of sea-pen, as a future scope, the analysis of pressure and local torque distribution on novel blade should be thoroughly studied in order to know whether or not the low-pressure region or attached vortices are strengthened. In addition to that, the entire power characteristics of the novel blade and literature supporting the aforementioned analogy should be worth exploring.

References

1. J. Sarma, S. Jain, P. Mukherjee, U.K. Saha, Hybrid/combined Darrieus-Savonius wind turbines: erstwhile development and future prognosis. J. Sol. Energy Eng. Trans. ASME. **143**, 1–20 (2021). https://doi.org/10.1115/1.4050595
2. N. Alom, U.K. Saha, Four decades of research into the augmentation techniques of Savonius wind turbine rotor. J. Energy Resour. Technol. **140**, 050801 (2018). https://doi.org/10.1115/1.4038785
3. S. Roy, U.K. Saha, Review of experimental investigations into the design, performance and optimization of the Savonius rotor. Proc. Inst. Mech. Eng. Part A J. Power Energy. **227**, 528–542 (2013). https://doi.org/10.1177/0957650913480992
4. U.H. Rathod, P.K. Talukdar, V. Kulkarni, U.K. Saha, Effect of capped vents on torque distribution of a semicircular-bladed Savonius wind rotor. ASME J. Energy Resour. Technol. **141**, 101201 (2019). https://doi.org/10.1115/1.4043791
5. C.M. Chan, H.L. Bai, D.Q. He, Blade shape optimization of the Savonius wind turbine using a genetic algorithm. Appl. Energy. **213**, 148–157 (2018). https://doi.org/10.1016/j.apenergy.2018.01.029
6. A. Ramadan, K. Yousef, M. Said, M.H. Mohamed, Shape optimization and experimental validation of a drag vertical axis wind turbine. Energy **151**, 839–853 (2018). https://doi.org/10.1016/j.energy.2018.03.117
7. W. Tian, B. Song, J.H. Van Zwieten, P. Pyakurel, Computational fluid dynamics prediction of a modified savonius wind turbine with novel blade shapes. Energies **8**, 7915–7929 (2015). https://doi.org/10.3390/en8087915
8. C. Ma, L. Song, M.Z. Zhang, Performance study for a novel vertical axis wind turbine based on simulation analysis, in *Proceedings 2017 IEEE 14th International Conference Networking, Sensing and Control, ICNSC 2017*, pp. 549–554 (2017). https://doi.org/10.1109/ICNSC.2017.8000151
9. N. Fujisawa, F. Gotoh, Visualization study of the flow in and around a Savonius rotor. Exp. Fluids. **12**, 407–412 (1992). https://doi.org/10.1007/BF00193888
10. N. Fujisawa, On the torque mechanism of Savonius rotors. J. Wind Eng. Ind. Aerodyn. **40**, 277–292 (1992). https://doi.org/10.1016/0167-6105(92)90380-S
11. V. Modi, M. Fernando, On the performance of the Savonius wind turbine. ASME J. Sol. Energy Eng. **111**, 71–81 (1989). https://doi.org/10.1115/1.3268289
12. C. Kailash, C. Raghunathan, *Faunal Diversity of Biogeographic Zones: Islands of India.* Director, Zoological Survey of India, Kolkata, India (2018)
13. G.F. Warner, On the shape of passive suspension feeders, in *Biology of Benthic Organisms*, ed. by B.F. Keegan, P.O. Ceidigh, P.J.S. Boaden (Pergamon Press Ltd, 1977), pp. 567–576
14. M.A.R. Koehl, Water flow and the morphology of Zoanthid colonies, in *Proceedings, Third International Coral Reef Symposium.* Rosensteil School of Marine and Atmospheric Science, University of Miami, Miami, Florida, pp. 437–444 (1977)
15. G.J. Leversee, Flow and feeding in fan-shaped colonies of the gorgonian coral, Leptogorgia. Biol. Bull. **151**, 344–356 (1976). 1540667
16. J. Andrews, Orange sea-pen (Ptilosarcus gurneyi) (CC-BY-NC), https://www.inaturalist.org/observations/51382128
17. S. Vogel, *The Life in Moving Fluid—The Physical Biology of Flow* (Princeton University Press, Princeton, New Jersey, 1994)
18. B.A. Best, Passive suspension feeding in a sea pen: effects of ambient flow on volume flow rate and filtering efficiency. Biol. Bull. **175**, 332–342 (1988). 1541723
19. R.L. Shimek, To pen a tale of Pens. https://ibis.geog.ubc.ca/biodiversity/efauna/documents/ToPenATaleOfPens.pdf
20. R. Kathleen, Orange Sea-Pen (Ptilosarcus gurneyi) (CC-BY-NC), https://doi.org/10.15468/dl.redsde. https://www.inaturalist.org/observations/66457417
21. MATLAB Version (R2021a). The MathWorks Inc., Natick, Massachusetts (2021)

22. ANSYS FLUENT 14.5. Ansys Inc., Canonsburg, PA 15317 USA (2012)
23. J. Sun, B. Bhushan, The structure and mechanical properties of dragonfly wings and their role on flyability. Comptes Rendus—Mec. **340**, 3–17 (2012). https://doi.org/10.1016/j.crme.2011.11.003
24. D. Satrio, I.K.A.P. Utama, Mukhtasor, Numerical investigation of contra rotating vertical-axis tidal-current turbine. J. Mar. Sci. Appl. **17**, 208–215 (2018). https://doi.org/10.1007/s11804-018-0017-5
25. K.B. Lua, H. Lu, T.T. Lim, Rotating elliptic cylinders in a uniform cross flow. J. Fluids Struct. **78**, 36–51 (2018). https://doi.org/10.1016/j.jfluidstructs.2017.12.023
26. G. Bangga, A. Dessoky, Z. Wu, K. Rogowski, M.O.L. Hansen, Accuracy and consistency of CFD and engineering models for simulating vertical axis wind turbine loads. Energy **206**, 118087 (2020). https://doi.org/10.1016/j.energy.2020.118087
27. N. Alom, U.K. Saha, A. Dewan, In the quest of an appropriate turbulence model for analyzing the aerodynamics of a conventional Savonius (S-type) wind rotor. J. Renew. Sustain. Energy. **13** (2021). https://doi.org/10.1063/5.0034362
28. J.M. Edwards, L. Angelo Danao, R.J. Howell, Novel experimental power curve determination and computational methods for the performance analysis of vertical axis wind turbines. ASME J. Sol. Energy Eng. **134**, 031008 (2012). https://doi.org/10.1115/1.4006196
29. ANSYS Fluent User's Guide. ANSYS, Inc., Canonsburg, PA 15317 USA (2011)
30. N. Alom, U.K. Saha, Evolution and progress in the development of savonius wind turbine rotor blade profiles and shapes. J. Sol. Energy Eng. Trans. ASME. **141**, 030801 (2019). https://doi.org/10.1115/1.4041848

Study of Gas–Liquid Flow in a Curved Microchannel for Sustainable Energy Application

Deepak Kumar Mishra, Anugrah Singh, and Raghvendra Gupta

Abstract Multiphase flow curved microchannels have shown enormous applicability in sustainable energy applications such as compact heat exchanger, process intensification, and solar thermal energy storage. This study investigates gas–liquid flow regimes in 1 mm diameter, circular, glass microchannel having U-bends. Ethylene glycol mixture (50% water) and nitrogen are used as liquid and gas phases. Images of various flow regimes are captured using a high-speed camera with high-intensity LED light. Slug, slug-annular, annular, and churn flow regimes are observed. Bubbles in slug flow regime, move closer to the inner wall of the bend. High-amplitude waves stand at the bend and merge with the next wave in slug-annular flow. However, a new high-amplitude wave shows up after crossing the U-bend. The gas core in annular flow regime moves toward the outer wall of the bend, and only the inner interface contains small ripples. In churn flow, interfaces are observed to be wavy at the curvature. A typical flow regime map is developed on gas and liquid Dean number coordinate ($\mathrm{De_G} - \mathrm{De_L}$).

Keywords Curved channel flow · Flow regime map · Slug flow · Annular flow · Slug-annular flow · Flow visualization

Abbreviations

U_S	Superficial velocity (m/s)
Re	Reynolds number $\left(\frac{dV\rho}{\mu}\right)$ (–)
Ca	Capillary number $\left(\frac{\mu V}{\sigma}\right)$ (–)
We	Weber number $\left(\frac{\rho D V^2}{\sigma}\right)$ (–)
Su	Suratman number $\left(\frac{d\rho_L \sigma}{\mu_l^2}\right)$ (–)

D. K. Mishra (✉) · A. Singh · R. Gupta
Department of Chemical Engineering, Indian Institute of Technology Guwahati, Guwahati, Assam 781039, India
e-mail: deepak.mishra@iitg.ac.in

V. S. Moholkar et al. (eds.), *Sustainable Energy Generation and Storage*,
https://doi.org/10.1007/978-981-99-2088-4_2

Oh Ohnesorge number $\left(\frac{\mu_l}{\sqrt{2\rho_l \sigma R}}\right)$ (–)
De Dean number $\left(\mathrm{De} = \mathrm{Re}\sqrt{\frac{D}{2R_c}}\right)$ (–)
Q Flow rate (m^3/s)
ρ Density (kg/m^3)
V Average velocity (m/s)
μ Dynamic viscosity (Pa s)
σ Surface tension (N/m)
R Radius
L Characteristic length (m)

Subscript

l Liquid phase
SG Superficial gas
SL Superficial liquid

1 Introduction

Gas–liquid flow in microchannels has shown tremendous application possibilities in sustainable energy [1]. Due to high interfacial area density, enhanced heat and mass transfer rates can be achieved in curved microchannels [2]. Major applications such as active microchannel cooling of concentrator photovoltaics [3], gas–liquid flow in PEM electrolyzer [4, 5], require an understanding of flow regimes in a microchannel for better design and optimization [6]. A number of U bends are frequently used to increase the channel length in order to maintain the longer residence time of the fluids in the channel. Based on the gas and liquid flow rates, various flow regimes such as bubbly, slug, annular, and churn may occur in the channel [7]. The curvature of the bend introduces secondary flow or Dean vortices in the flow [8], which affect the flow profile and interface behavior. Therefore, it is important to understand the hydrodynamics of gas–liquid flow regimes in the U-bend section of the channel.

Günther et al. [9] investigated flow using microparticle image velocimetry in straight and meandering channels for gas–liquid chemical reaction applications. In comparison to straight microchannels meandering channel enhances mixing for chemical reaction with a residence time of several minutes. Fries and Rohr [10] studied mass transfer in a rectangular meandering microchannel for gas–liquid flow. They showed that mixing inside liquid slug occurs, which in turn enhances mass transfer. Further, it is suggested to investigate large Dean number $\left(\mathrm{De} = \mathrm{Re}\sqrt{\frac{D}{2R_c}}\right)$ flow for mixing intensification. Roudet et al. [11] investigated air–water two-phase flow in straight and meandering microchannels and showed that curvature delays the transition of flow patterns from slug to slug-annular flow regime. Bubble length increases due to curvature while liquid slug length remains almost of the same size.

Sarkar et al. [12] studied liquid–liquid flow in serpentine microchannel. They observed seven flow patterns and developed a flow regime map on flow rate coordinates. Wu and Sundén [13] investigated liquid–liquid flow regimes in straight and serpentine microchannels and showed the effect of the bend on flow patterns. They compared the centrifugal forces and analyzed that bend affects the core phase by breaking it into slugs and droplets.

Table 1 shows some important parameters which play significant role in flow regime distribution at the curved microchannel.

Studies of gas–liquid flow patterns in the meandering channel have shown significant application in heat and mass transfer processes along with reaction and mixing. Although there is a lot of research on flow regime distribution in meandering microchannels for liquid–liquid flow, there is still need for a gas–liquid flow regime map in curving microchannels. Therefore, flow regime map of gas–liquid flow in curved microchannel becomes important as it suggests the basic flow condition. It is important to know the experimental condition of flow regimes to apply for the reaction, heat, and mass transfer applications. Therefore, this study focuses on the gas–liquid flow patterns in U-bend and development of a flow regime map.

Table 1 Some important parameter selected by various authors in the literature for flow regime map

Parameter	Significance	References
U_S	Superficial velocity	Triplett et al. [7], Patel et al. [14], Fukano and Kariyasaki [15], Kawaji and Chung [16], Günther et al. [9]
Re	Inertial force to viscous force	Dessimoz et al. [17]
Ca	Viscous force to surface tension force	Cubaud and Mason [18]
We = Re × Ca	Inertial force to surface tension force	Akbar et al. [19]
Su = Re/Ca	Surface tension to momentum transport inside the fluid	Jayawardena et al. [20]
Oh = $\sqrt{}$We/Re	Viscous to inertia and surface tension	Wu and Sundén [13]
$Q_l/(Q_l + Q_g)$	Flow rate	Suo and Griffith [21]
ρV^2	Inertial force	Hewitt and Roberts [22]

2 Methodology

High-speed flow visualization experiments are performed on gas–liquid flow in 1 mm diameter circular cross-sectional microchannel having U-bends. Glass channel (procured from LTF Germany) consists of multiple straight sections connected with U-bends. Nitrogen gas is used as a gas phase controlled by a mass flow controller (Alicat MC 5SLPM). Ethylene glycol mixture with de-ionized water (50%) is used as the liquid phase. Measured fluid properties are shown in Table 2.

The liquid phase flow rate is controlled by a syringe pump (Holmarc HO SPLF2). Images of flow patterns are captured by a high-speed camera (PHOTRON, Mini UX50) using a flicker-free LED light (YONGNUO YN900). A schematic diagram of the experimental setup is shown in Fig. 1. Microchannel consists of 21 U-bend and 20 straight channels. The dashed rectangle region on the microchannel is enlarged in Fig. 2.

Figure 2 shows the selected U-bend region where flow regimes are studied. To ensure that no inlet or exit effect occurs, a U-bend positioned in the middle position from the inlet and outlet is used for flow observation. The radius of curvature of the center line of the bend is 1 mm.

Table 2 Fluid properties used in the experiments

Fluids	Density (ρ) (kg/m^3)	Viscosity (μ) (Pa. S)	Surface tension (σ) (N/m)
50% Ethylene Glycol + water	1055	0.004	0.057
Nitrogen (N_2)	1.145	0.00002	–

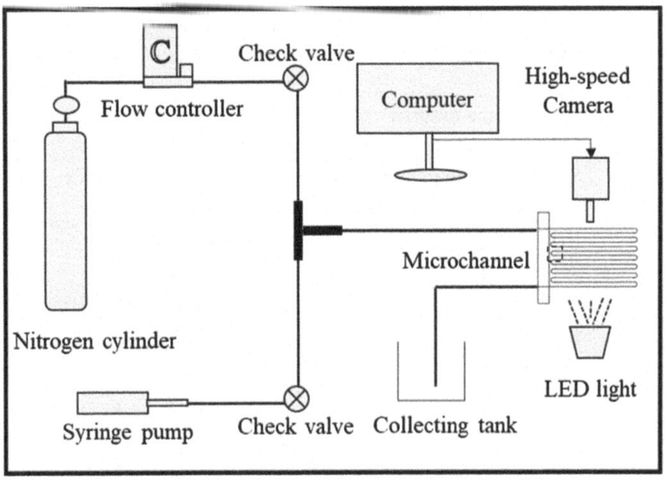

Fig. 1 Schematic diagram of the experimental setup

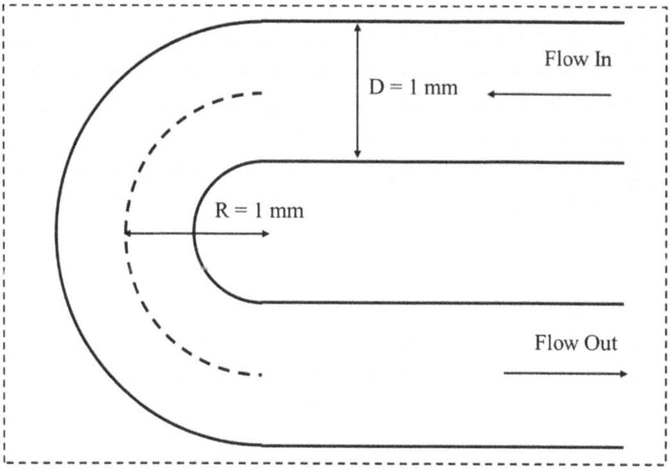

Fig. 2 Enlarged view of U-bend showing the flow inlet and outlet direction

3 Results

Experiments are performed over a large range of gas and liquid superficial velocities. Gas superficial velocity (U_{SG}) varies from 2.34 to 46.83 m/s, and liquid superficial velocity varies from 0.01 to 0.20 m/s. In this range of the flow rates, four flow regimes are observed which are slug flow, slug-annular flow, annular flow, and churn flow regimes. Notably, the bubbly flow is not observed in the present work as the gas flow rate is high.

3.1 Slug Flow Regime

At low gas superficial velocity and low liquid superficial velocity, slug flow regime is observed. Images of bubble evolution with time at the curvature are shown in Fig. 3. Note that observed bubbles are long; hence, entire bubble is not captured in one frame. At lowest liquid superficial velocity, liquid slug is not observed between two bubbles Fig. 1a. Bubble and liquid slug both are longer in this case. An increase in U_{SG} (Fig. 3b) at constant U_{SL} results in the reduction of the liquid slug length. Further increase in U_{SL} at constant U_{SG} (Fig. 3c) introduces a long liquid slug between bubbles. Note that in the slug flow regime, bubble always moves near to the inner wall of the curvature.

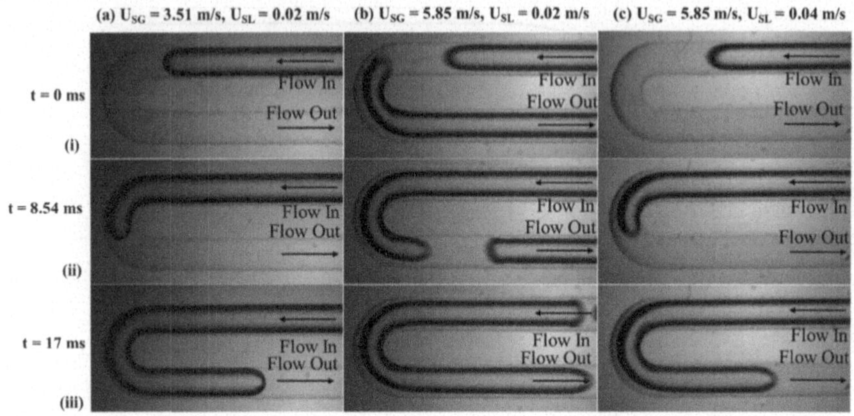

Fig. 3 Images of gas–liquid slug flow evolution with time at t = (i) 0 ms, (ii) 8.54 ms, and (iii) 17 ms at **a** $U_{SG} = 3.51$ m/s and $U_{SL} = 0.02$ m/s, **b** $U_{SG} = 5.85$ m/s and $U_{SL} = 0.02$ m/s, and **c** $U_{SG} = 5.85$ m/s and $U_{SL} = 0.04$ m/s

3.2 Slug-Annular Flow

Further, a decrease in the length of the liquid slug is caused by an increase in gas superficial velocity at a constant liquid superficial velocity (0.02 m/s). At moderate U_{SG}, continuous gas core with high amplitude waves is observed. Figure 4 shows a time evolution of slug-annular flow at $U_{SG} = 7.02$ and $U_{SL} = 0.02$ m/s. The behavior of two high-amplitude waves at the curvature, and their interaction is shown. The first high-amplitude wave is captured at $t = 0$ ms, which is about to pass through curvature. The initial wave was spotted standing at the curved region at $t = 35$ ms, while another high-amplitude wave approached curvature. At $t = 65$ ms, the second wave enters the curved region while the first wave is still observed to be standing. The first wave moves out of the curvature at $t = 95$ ms, but the second wave continues to stand in the same position as the first. At $t = 135$ ms, both waves show a decrease in amplitude but remain at the same spot while another high amplitude wave appears in the region of interest. At $t = 150$ ms, both waves merged and formed a high-amplitude wave, and this wave was observed to move further in the straight channel.

3.3 Annular Flow

Increasing gas superficial velocity at low liquid superficial velocity results in gas–liquid annular flow. Figure 5 shows gas–liquid annular flow at $U_{SG} = 39.8$ m/s, $U_{SL} = 0.01$ m/s. Very low amplitude waves are observed at the interface. The gas core moves toward the outer wall due to the high inertia of the gas phase, and liquid is displaced from the outer wall. A small liquid region is observed at the inner wall

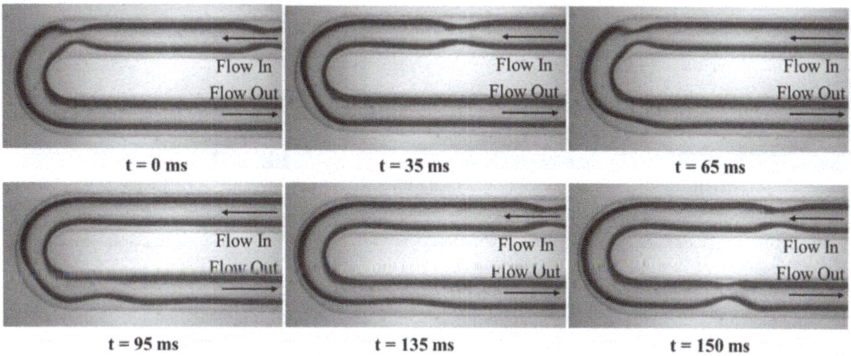

Fig. 4 Images of gas–liquid slug-annular flow evolution with time at $U_{SG} = 7.02$ m/s and $U_{SL} = 0.02$ m/s

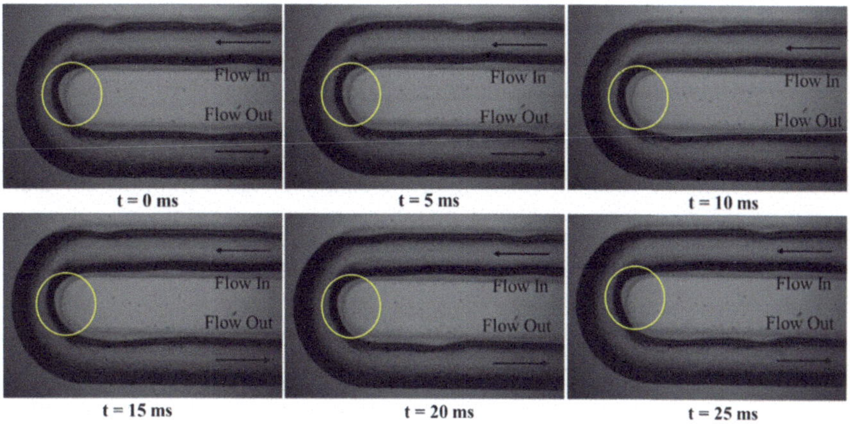

Fig. 5 Images of gas–liquid annular flow evolution with time at $U_{SG} = 39.8$ m/s and $U_{SL} = 0.01$ m/s

shown in the circle. This liquid zone's size is observed to change with the small wave entering into curvature. It is also observed that the gas core remains at the outer wall after exiting from curvature. Small waves reappear at the inner wall interface with the gas core propagation.

3.4 *Churn Flow*

High gas and liquid superficial velocities result into churn flow regime. Figure 6 shows the time evolution images of churn flow in the curved region. As churning at the interface (shown in dashed box) passes through the curvature, the outer interface

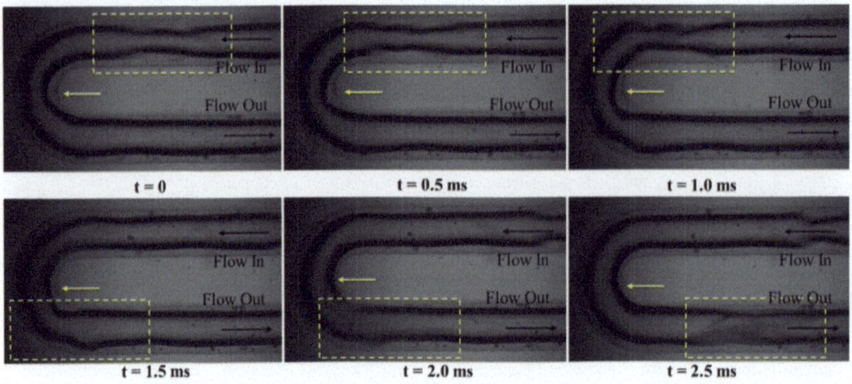

Fig. 6 Images of gas–liquid churn flow evolution with time at $U_{SG} = 35.12$ m/s and $U_{SL} = 0.08$ m/s

becomes wavy, unlike the annular flow regime. High inertial force of liquid phase moves toward outer wall resulting both interfaces wavy at the curved region. The outer and inner interface becomes smoother (at $t = 2.5$ ms) as the churning area passes through the curvature. The entering churning region is observed to be having high wave amplitude while the exiting churning region dispersed throughout the cross-sectional area of the channel. Crossing of the churning region from the curvature also affects the inner and outer liquid film (observed at the mid of the curved region at $t = 0$–2.5 ms).

3.5 *Flow Regime Map*

Figure 7 shows a comparison between flow regimes observed in the U-bend experiments with flow regimes obtained by Triplett et al. [7] for air–water flow in 1.09 mm diameter straight channel. Dashed line shown on the map is just to differentiate regime transition. Bubbly flow regime is not observed due to high-velocity range used in the experiment.

Significant difference in the flow regimes and regime transition is observed between this experimental study and data obtained by Triplett et al. [7]. This difference can be attributed to the curvature of the U-bend.

A flow regime map is plotted on the coordinate of Dean number of gas and liquid phase ($De_G - De_L$) and shown in Fig. 8a. Four flow regimes are observed on the superficial velocity range available for experiments, namely slug, slug-annular, annular, and churn flow regimes (Fig. 8b). Note that the dotted lines shown on the map are only used for the differentiation of flow regimes and do not show any critical values of regime transition. At low De_G and De_L slug flow is observed, whereas at very high values, churn flow regime appears. Increasing De_G at constant De_L results in annular flow transition.

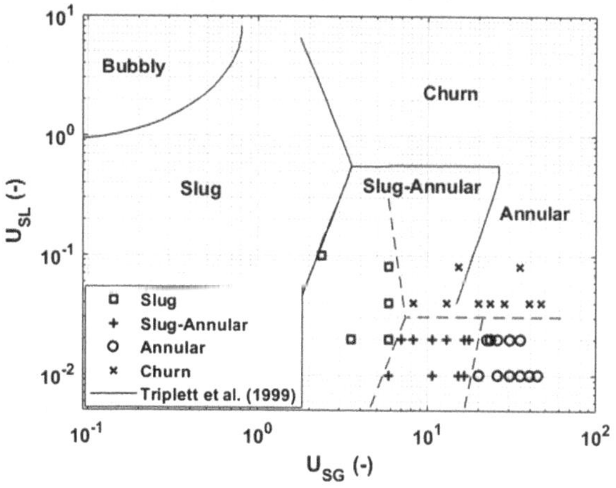

Fig. 7 Flow regime map for 1 mm diameter circular cross-sectional U-bend superimposed on straight channel flow regime data by Triplett et al. [7]

4 Conclusion

Effect of 1 mm diameter circular cross-sectional U-bend/curved channel on gas–liquid flow is studied using a high-speed camera. Based on the superficial velocity range of gas and liquid, four flow regimes are observed, namely slug flow, slug-annular flow, annular flow, and churn flow at the curvature. The liquid film is observed to be non-symmetric on the outer and inner walls because of curvature. In the slug flow regime, an increase in U_{SG} introduces smaller bubbles and shorter liquid slugs, while an increase in U_{SL} introduces long bubbles with bigger liquid slugs. In the slug-annular flow regime, multiple high-amplitude waves are observed, which merge at the U-bend due to the curvature effect. However, high-amplitude waves reappear after passing through the bend. In the annular flow regime, small ripples at the interface are observed, which turn into a liquid zone near the inner wall at the curvature while passing through the bend. The gas core always remains near the outer wall in the annular flow regime. The curvature also affects the churn flow by enhancing the churning at the interface. Both outer and inner interfaces become wavy while passing through the bend. Based on the experimental observation, a flow regime map is developed on gas and liquid Dean number coordinate. In the future, a parametric study of fluid properties and channel curvature radius can be done for an in-depth understanding of the curvature effect on flow regimes.

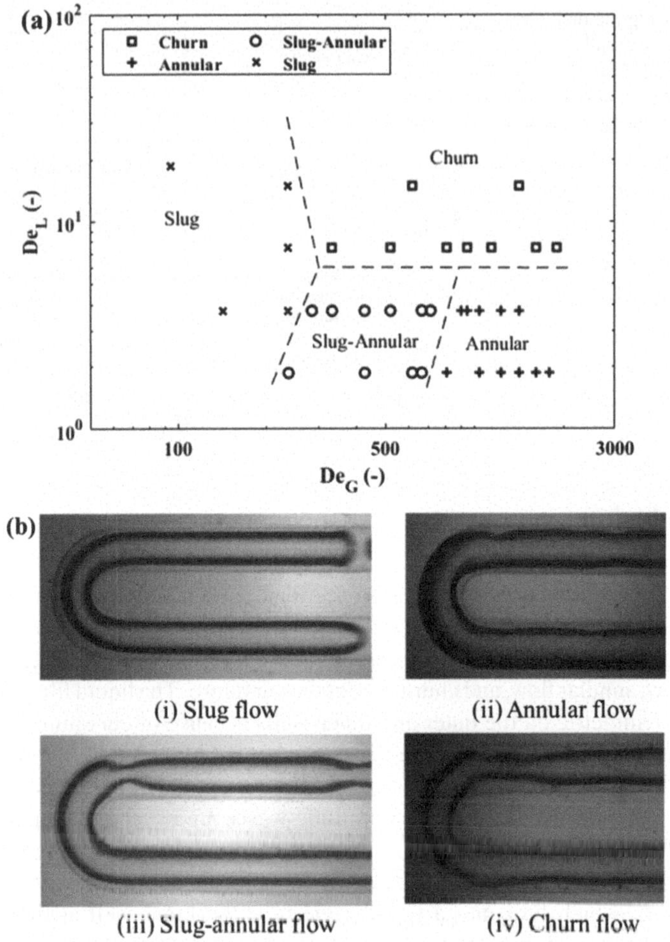

Fig. 8 **a** Flow regime map showing the observed gas–liquid flow regimes in 1 mm circular diameter U-bend. Dotted lines on the map just differentiate different flow regimes and do not show any critical regime transition value. **b** Instantaneous snapshot of corresponding flow regimes: (i) slug flow, (ii) annular flow, (iii) slug-annular flow, and (iv) churn flow

Acknowledgements We acknowledge the funding support provided by the Science and Engineering Research Board (SERB), Department of Science and Technology (DST), Government of India, under the Young Scientist Fast Track Scheme (No: SB/FTP/ETA/0382-2013).

References

1. J. Čejková, T. Banno, M.M. Hanczyc, F. Štěpánek, Droplets as liquid robots. Artif. Life **23**, 528–549 (2017)
2. R. Gupta, D. Fletcher, B. Haynes, Taylor flow in microchannels: a review of experimental and computational work. J. Comput. Multiphase Flows **2**, 1–31 (2010)
3. N. Gilmore, V. Timchenko, C. Menictas, Microchannel cooling of concentrator photovoltaics: a review (2018)
4. S.S. Lafmejani, A.C. Olesen, S.K. Kær, VOF modelling of gas–liquid flow in PEM water electrolysis cell micro-channels. Int. J. Hydrogen Energy **42**, 16333–16344 (2017)
5. S. Sadeghi Lafmejani, A.C. Olesen, S. al Shakhshir, S.K. Kær, Analysing gas-liquid flow in PEM electrolyser micro-channels using a micro-porous ceramic as gas permeable wall. ECS Trans. **80**, 1107–1115 (2017)
6. T. Xiong, G. Liu, S. Huang, G. Yan, J. Yu, Two-phase flow distribution in parallel flow mini/micro-channel heat exchangers for refrigeration and heat pump systems: a comprehensive review (2022)
7. K.A. Triplett, S.M. Ghiaasiaan, S.I. Abdel-Khalik, D.L. Sadowski, Gas-liquid two-phase flow in microchannels part I: two-phase flow patterns. Int. J. Multiph. Flow **25**, 377–394 (1999)
8. W.R. Dean, J.M. Hurst, Note on the motion of fluid in a curved pipe. Mathematika **6**, 77–85 (1959)
9. A. Günther, S.A. Khan, M. Thalmann, F. Trachsel, K.F. Jensen, Transport and reaction in microscale segmented gas-liquid flow. Lab Chip **4**, 278–286 (2004)
10. D.M. Fries, P. Rudolf von Rohr, Liquid mixing in gas-liquid two-phase flow by meandering microchannels. Chem. Eng. Sci. **64**, 1326–1335 (2009)
11. M. Roudet, K. Loubiere, C. Gourdon, M. Cabassud, Hydrodynamic and mass transfer in inertial gas-liquid flow regimes through straight and meandering millimetric square channels. Chem. Eng. Sci. **66**, 2974–2990 (2011)
12. P.S. Sarkar, K.K. Singh, K.T. Shenoy, A. Sinha, H. Rao, S.K. Ghosh, Liquid-liquid two-phase flow patterns in a serpentine microchannel. Ind. Eng. Chem. Res. **51**, 5056–5066 (2012)
13. Z. Wu, B. Sundén, Liquid-liquid two-phase flow patterns in ultra-shallow straight and serpentine microchannels. Heat Mass Transf. **55**, 1095–1108 (2019)
14. R.S. Patel, J.A. Weibel, S.V. Garimella, Characterization of liquid film thickness in slug-regime microchannel flows. Int. J. Heat Mass Transfer. **115**, 1137–1143 (2017)
15. T. Fukano, A. Kariyasaki, Characteristics of gas-liquid two-phase flow in a capillary tube (1993)
16. P.M.Y. Chung, M. Kawaji, The effect of channel diameter on adiabatic two-phase flow characteristics in microchannels. Int. J. Multiph. Flow **30**, 735–761 (2004)
17. A.L. Dessimoz, P. Raspail, C. Berguerand, L. Kiwi-Minsker, Quantitative criteria to define flow patterns in micro-capillaries. Chem. Eng. J. **160**, 882–890 (2010)
18. T. Cubaud, C.M. Ho, Transport of bubbles in square microchannels. Phys. Fluids **16**, 4575–4585 (2004)
19. M.K. Akbar, D.A. Plummer, S.M. Ghiaasiaan, On gas–liquid two-phase flow regimes in microchannels. Int. J. Multiph. Flow **29**, 855–865 (2003)
20. S.S. Jayawardena, V. Balakotaiah, L.C. Witte, Flow pattern transition maps for microgravity two-phase flows. AIChE J. **43**, 1637–1640 (1997)
21. M. Suo, P. Griffith, Two-phase flow in capillary tubes. J. Basic Eng. **86**, 576–582 (1964)
22. G.F. Hewitt, D.N. Roberts, Studies of two-phase flow patterns by simultaneous x-ray and flash photography. Aere-M 2159 (1969)

Design of a Microreactor for Biodiesel Synthesis

Mohammad Anzar Hussain and Raghvendra Gupta

Abstract Biodiesel is a sustainable alternative energy source to petroleum-based sources; it is renewable and environment friendly. Yet, in terms of cost, it still has a long way to go to compete with conventional fossil fuels. Biodiesel is generally produced using transesterification reaction and several lab-scale studies exist for the generation of biodiesel using stirred tank reactors operated in batch mode. The yield of biodiesel is limited by the mass transfer between immiscible oil and methanol phases. The mixing and continuous production of biodiesel can be significantly enhanced using microreactors (reactors having hydraulic diameter $\leq$ 1 mm), due to the fact that as the reactor size decreases, the surface-to-volume ratio increases, which leads to increase in net interfacial area available for mass transfer. The flow inside microreactors is predominantly laminar flow. As compared to turbulent flow, laminar flow provides excellent controllability, but at the cost of poor mixing. The mixing can be further enhanced by making the microreactor flow path curved, allowing the secondary flow induced mixing to happen. In addition to the shape of microreactor, the type of multiphase flow regime also plays an important role in determining the distribution of phases in each other. Combining these features can significantly enhance the yield of a mass transfer limited reaction such as transesterification reaction. Hence, we perform a reactor design calculation based on available relevant data in literature to design a microreactor for producing biodiesel from the transesterification of triglycerides in soybean oil with methanol using a homogeneous catalyst. The result of the calculations is then compared with the experimental data available by other researchers.

Keywords Transesterification · Biodiesel · Plug flow microreactor

M. A. Hussain · R. Gupta (✉)
Department of Chemical Engineering, Indian Institute of Technology Guwahati, Guwahati, Assam 781039, India
e-mail: guptar@iitg.ac.in

R. Gupta
Center for Sustainable Polymers, Indian Institute of Technology Guwahati, Guwahati, Assam 781039, India

V. S. Moholkar et al. (eds.), *Sustainable Energy Generation and Storage*,
https://doi.org/10.1007/978-981-99-2088-4_3

1 Introduction

Biodiesel is a sustainable alternative energy source to petroleum-based sources; it is renewable and environment friendly. Biodiesel is derived from vegetable oils, waste kitchen frying oils, or animal fats via the transesterification reaction of triglyceride with methanol in the presence of a homogeneous catalyst like NaOH (alkali-catalyzed transesterification reaction) and H_2SO_4 (acid-catalyzed transesterification reaction). The alkali-catalyzed transesterification reaction relies heavily on mass transfer at the interface between both the phases, as the reactants are immiscible liquids and the interfacial area between both the phases plays a very important role in mass transfer [1, 2]. As the reactor size decreases, the surface-to-volume ratio increases, leading to an increase in interfacial area density and mass transfer. For this reason, microreactors are often proposed for increasing the yield of the transesterification reaction. However, due to the decrease in size, the flow in microreactors is generally laminar, which leads to poor mixing. Mixing enhancement in microfluidic channels can be done in two ways: active and passive mixing [3]. Passive mixing microfluidic systems are generally desirable since they do not involve any moving parts and hence no requirement of extra power source for mixing. One way to achieve passive mixing in microreactors is by making the flow path curved, to generate the secondary flow induced mixing caused by Dean vortices. Hence, using helical microreactors for the production of biodiesel can result in better transesterification reaction rates, yield and continuous production. Another approach to passive mixing is by choosing an appropriate flow regime for the two immiscible reactant phases, as the flow behavior in two-phase flow is regime dependent. For example, the Taylor or slug flow regime involves long, regular droplets of one fluid, having diameter almost same as that of the channel diameter, in a continuous liquid stream. These droplets are separated by slugs of continuous liquid [4]. The droplets of the dispersed phase and slug of continuous phase have recirculation motion within them resulting in vigorous mixing in each phase. In addition to it, the large interfacial area achieved between two phases can give rise to very high rates of interfacial mass transfer. The current work attempts to design and conceptualize a helical microreactor for the continuous synthesis of biodiesel.

In this chapter, we aim to give a reactor design calculation for a microreactor idealized as a plug flow reactor for the continuous synthesis of biodiesel. The alkali-catalyzed synthesis of biodiesel is planned to be carried out in a capillary helical microreactor having inner diameter of 1 mm with soybean oil and methanol as reactants. The soybean oil and NaOH-methanol solutions get to mix in a T-junction with molar ratios 1:6, followed by passing of the reacting mixture through the helical microreactor (see Fig. 1) at 323.15 K temperature maintained via a heat bath. The initial flow rate of triglycerides phase is to be fixed at 4.2 cm^3/h with initial concentration of triglycerides phase at 0.774 mol/L. Based on such requirements, the calculated design specifications of the helical microreactor are to be calculated. The material of construction chosen is stainless steel; however, to identify the type of flow regime,

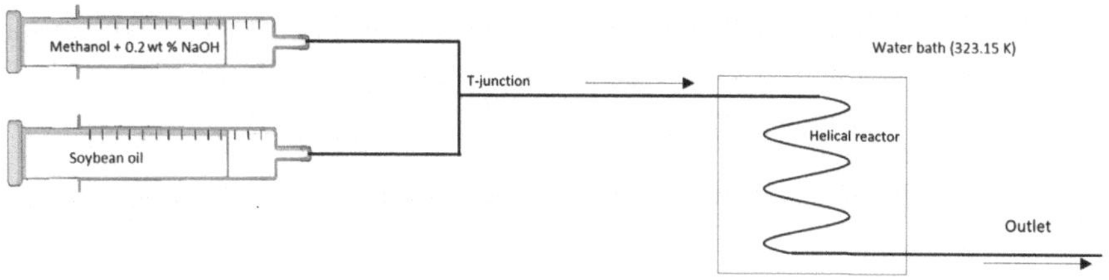

Fig. 1 Schematic of the overall process

transparent plastic tubing is to be placed at both inlet and outlet of the helical reactor design.

Although there are many studies available [5] on the kinetics of transesterification reaction of esters with alcohol, very few of these take account of the variations in mass transfer limitations on using microreactors. For our initial design calculation, we use the kinetic study of Noureddini and Zhu conducted at a stirred tank reactor for a rotational speed of 300 rpm [1], to calculate the length of the flow microreactor by assuming it to provide same level of mixing intensity. For the multiphase reactions, such as transesterification reaction, the degree of mixing plays a very important role, as soybean oil and methanol are both immiscible. So a change in the mixing intensity is expected to affect the transesterification reaction kinetics. The effect of mixing intensity was studied by applying mixing using a mechanical stirrer. Noureddini and Zhu [1] represented the degree of mixing in terms of Reynolds number (N_{Re}). The Reynolds number and impeller speed are related as:

$$N_{\text{Re}} = \frac{n D_a^2 \rho}{\mu} \tag{1}$$

where n implies impeller rotational speed, D_a implies impeller diameter, μ and ρ are fluid mixture viscosity and density. The kinetic data that we employ in this work are reported at 50 °C and $N_{\text{Re}} = 6200$ [1].

2 Problem Formulation

The entire process involves transesterification of triglycerides (TG) present in vegetable oil with methanol (CH_3OH), in presence of an alkaline catalyst (NaOH) to yield methyl esters of fatty acids ($RCOOCH_3$) also called as fatty acid methyl esters (FAME) and glycerol (GL). The intermediates formed are di-glycerides (DG) and monoglycerides (MG). The reactants are immiscible and form a two-phase liquid system. However, the product methyl esters act as a mutual solvent for the reactants and form a single phase system. Hence for simplicity, we assume the system to be a single-phase system. In the current study, to perform the design calculation of a continuous flow microreactor, we undertake the following assumptions:

i. The reactant mixture is assumed to be a homogeneous mixture phase with molar ratio of alcohol to triglycerides 1:6 at the T-junction.
ii. The helical microreactor to be designed is idealized as a straight plug flow reactor of 1 mm inner diameter and 3.14 mm^2 cross-sectional area (a).
iii. Quasi-steady-state approximation to be valid for the intermediate species DG and MG.

From the process design calculations, reactor volume is obtained which translates to the determination of reactor length (l) for given reactor diameter and for a given value of conversion of triglycerides. The transesterification reaction is a second order reaction; the steps of its mechanism are given by Eqs. (2–4):

$$TG + CH_3OH \underset{k_2}{\overset{k_1}{\rightleftharpoons}} DG + R_1COOCH_3 \tag{2}$$

$$DG + CH_3OH \underset{k_4}{\overset{k_3}{\rightleftharpoons}} MG + R_2COOCH_3 \tag{3}$$

$$MG + CH_3OH \underset{k_6}{\overset{k_5}{\rightleftharpoons}} GL + R_3COOCH_3 \tag{4}$$

That makes the overall reaction as:

$$TG + 3CH_3OH \underset{k_8}{\overset{k_7}{\rightleftharpoons}} GL + 3RCOOCH_3 \tag{5}$$

In current study for the design calculation of the microreactor length, we evaluate the second order mechanism without the overall reaction (Eq. 5), because the values for k_7 and k_8 are negligible as compared to rate constants from k_1 to k_6 (see Table 1). The rate constants employed in the current study were obtained from the experimental data values of the alkali-catalyzed transesterification reaction carried out in a stirred batch reactor at 50 °C and $N_{Re} = 6200$ [1]. The reported rate constants are [1]:

Table 1 Average rate constants at 50 °C and $N_{Re} = 6200$

Rate constant	Reported values at 50 °C and $N_{Re} = 6200$
k_1	0.05 $L.(mol.min)^{-1}$
k_2	0.11 $L.(mol.min)^{-1}$
k_3	0.215 $L.(mol.min)^{-1}$
k_4	1.228 $L.(mol.min)^{-1}$
k_5	0.242 $L.(mol.min)^{-1}$
k_6	0.007 $L.(mol.min)^{-1}$
k_7	7.84×10^{-5} $L.(mol.min)^{-1}$
k_8	1.58×10^{-5} $L.(mol.min)^{-1}$

3 Mathematical Model

The system of governing equations is generated from the rate of elimination/generation of different species. The rate of generation/elimination of the triglycerides (TG) (Eq. 6), di-glycerides (DG) (Eq. 7), monoglycerides (MG) (Eq. 8), fatty acid methyl esters (E) (Eq. 9), methanol (A) (Eq. 9) and glycerol (GL) (Eq. 10) from the set of stepwise reactions (Eqs. 2–5) is given as:

$$\frac{\mathrm{d[TG]}}{\mathrm{d}t} = -k_1[\mathrm{TG}][\mathrm{A}] + k_2[\mathrm{DG}][\mathrm{E}]. \tag{6}$$

$$\frac{\mathrm{d[DG]}}{\mathrm{d}t} = k_1[\mathrm{TG}][\mathrm{A}] - k_2[\mathrm{DG}][\mathrm{E}] - k_3[\mathrm{DG}][\mathrm{A}] + k_4[\mathrm{MG}][\mathrm{E}] \tag{7}$$

$$\frac{\mathrm{d[MG]}}{\mathrm{d}t} = -k_4[\mathrm{MG}][\mathrm{E}] - k_5[\mathrm{MG}][\mathrm{A}] + k_6[\mathrm{GL}][\mathrm{E}] + k_3[\mathrm{DG}][\mathrm{A}] \tag{8}$$

$$-\frac{\mathrm{d[A]}}{\mathrm{d}t} = \frac{\mathrm{d[E]}}{\mathrm{d}t} = k_1[\mathrm{TG}][\mathrm{A}] - k_2[\mathrm{DG}][\mathrm{E}] + k_3[\mathrm{DG}][\mathrm{A}] - k_4[\mathrm{MG}][\mathrm{E}] + k_5[\mathrm{MG}][\mathrm{A}] - k_6[\mathrm{GL}][\mathrm{E}] \tag{9}$$

$$\frac{\mathrm{d[GL]}}{\mathrm{d}t} = k_5[\mathrm{MG}][\mathrm{A}] - k_6[\mathrm{GL}][\mathrm{E}] \tag{10}$$

where [A] and [E] represent the concentrations of alcohol and FAME. For performing the design calculation of a plug flow microreactor, we use the following design equation:

$$\frac{V}{F_{\mathrm{TG}_0}} = \int_{X_{\mathrm{TG},0}}^{X_{\mathrm{TG}}} \frac{dX_{\mathrm{TG}}}{-r_{\mathrm{TG}}} = \frac{1}{[\mathrm{TG}]_0} \int_{[\mathrm{TG}]_0}^{[\mathrm{TG}]} \frac{d[\mathrm{TG}]}{r_{\mathrm{TG}}} \tag{11}$$

where X_{TG} refers to the conversion of triglycerides, $\left(X_{\mathrm{TG}} = \frac{[\mathrm{TG}]_0 - [\mathrm{TG}]}{[\mathrm{TG}]_0}\right)$. In the study conducted by Noureddini and Zhu, it can be clearly inferred from the data (see Fig. 1 of [1]), that it is safe to assume quasi-steady state approximation for the intermediate species di-glyceride and monoglyceride. This leads them to be related as:

$$[\mathrm{DG}] = \frac{k_1[\mathrm{TG}][\mathrm{A}] + k_4[\mathrm{MG}][\mathrm{E}]}{k_2[\mathrm{E}] + k_3[\mathrm{A}]} \tag{12}$$

$$[\mathrm{MG}] = \frac{k_3[\mathrm{DG}][\mathrm{A}] + k_6[\mathrm{GL}][\mathrm{E}]}{k_4[\mathrm{E}] + k_5[\mathrm{A}]} \tag{13}$$

From the stoichiometry of Eq. (5), we can write the relation between [A] and $[\mathrm{A}]_0$ as:

$$[A] = [A]_0 - 3\{[\mathrm{TG}]_0 - [\mathrm{TG}]\} \tag{14}$$

Substituting (12) and (13) in (6) and simplifying give the rate of the reaction of triglycerides as:

$$\frac{\mathrm{d}[\mathrm{TG}]}{\mathrm{d}t} = \{-k_1[\mathrm{TG}] + k_2\left[\frac{(k_1(k_4+k_5)[\mathrm{TG}]) + (k_4k_6X_{\mathrm{TG}}[\mathrm{TG}]_0)}{(k_2+k_3)(k_4+k_5)\left(1 - k_4k_5\{[\mathrm{A}]_0 - 3\{[\mathrm{TG}]_0 - [\mathrm{TG}]\}\}^2\right)}\right]\} \{[\mathrm{A}]_0 - 3\{[\mathrm{TG}]_0 - [\mathrm{TG}]\}\} \tag{15}$$

In the current study, the input data of reactant species are: molar ratio of alcohol to triglycerides (1:6), initial concentration of oil ($[\mathrm{TG}]_0 = 0.774$ mol/L) and initial concentration of methanol ($[CH_3OH]_0 = 4.64$ mol/L). As the molar ratio of alcohol to oil is kept constant at 6:1. Hence, soybean oil (triglyceride) is the limiting reactant. The initial total flow rate was fixed at 4.2 cm^3/h in order to compare the result with Guan et al. [6]. Simplifying the Eq. (15) will become of the form:

$$r_{[\mathrm{TG}]} = P - Q[\mathrm{TG}] - R[\mathrm{TG}]^2 \tag{16}$$

where P, Q and R are the constants corresponding to the rate constants and initial concentrations of triglycerides and methanol. Plugging in the reported values of rate constants by [1] in (15) and the mentioned initial concentrations of triglycerides ($[\mathrm{TG}]_0 = 0.774$ mol/L) and methanol ($[CH_3OH]_0 = 4.64$ mol/L), we get the values of the constants as; $P = 0.007954$, $Q = 0.02835$ and $R = 0.04964$. Substituting (16) in the design Eq. (11) gives the required reactor length as

$$l = \frac{F_{\mathrm{TG}_0}}{a[\mathrm{TG}]_0} \int_{[\mathrm{TG}]_0}^{[\mathrm{TG}]} \frac{d[\mathrm{TG}]}{P - Q[\mathrm{TG}] - R[\mathrm{TG}]^2} \tag{17}$$

4 Results

The design calculations are done to predict the minimum reactor length required, before going in for the microreactor fabrication for the continuous production of biodiesel in the fabricated microreactor. In current paper for 70% conversion of triglycerides ($[\mathrm{TG}] = 0.2322$ mol/L) with initial flow rate of 4.2 cm^3/h and initial concentration of triglycerides as 0.774 mol/L, the calculated plug flow microreactor volume (V) comes out to be 3802 mm^3. And, for a fixed diameter at 1 mm, the calculated microreactor length (l) comes out as 1210 mm long. On taking this length as the helical length to design a helical microreactor of 12 number of turns will give us a microreactor of 32.08 mm helical height and same helical diameter.

5 Conclusion

The design calculations done in this paper are limited by the basic physical assumptions under the conditions, that the helical microreactor is idealized as straight plug flow reactor, the reactants form a single phase and the application of quasi-steady state approximation for the intermediate species. The investigations of this paper form a very small contribution to the problem of enhancement of continuous production of biodiesel in microreactors, that is to perform design calculation for the reactor length. In this paper for the aforementioned conditions, the calculated plug flow microreactor volume (V) comes out to be 3802 mm^3 and the calculated microreactor length (l) comes out as 1210 mm long. A helical microreactor of this length as the helical length with 12 number of turns will have helical height same as its helical diameter of 32.08 mm. Comparing this reactor length with the experimental results of Guan et al. [6] in which they were able to obtain 78.3% conversion of triglycerides in a tubular microreactor of inner diameter 0.8 mm and length of 1000 mm using sunflower oil at higher reaction temperature of 60 °C, at oil to alcohol molar ratio of 1:4.6. We come to a conclusion that the calculated results although close, but still do not compare very well with the experimental results of Guan et al. [6]. Higher conversion with almost 20% difference in reactor length can be attributed to the type of oil used, higher reaction temperature, enhancement of mixing parameters due to unaccounted recirculation patterns at liquid slugs and occurrence of dean vortices at tube bends [4]. From a practical point of view, this confirms that indeed using a continuous flow microreactor leads to better conversion values for biodiesel production, even better than the theoretically calculated values obtained from batch reactor kinetics. These results imply that there is a huge gap in theory for reactive flow microreactor systems. Unavailability of chemical kinetics data and interfacial mass transfer coefficients for multiphase systems like the one dealt in this paper renders performing any sort of design calculations and predicts the reactor size as useless. Nevertheless, the calculations done here can be used to gauge preliminary microreactor size as a rough estimate of maximum required microreactor size.

References

1. H. Noureddini, D. Zhu, Kinetics of transesterification of soybean oil. JAOCS, J. Am. Oil Chemists' Soc. **74**(11) (1997).
2. E. Santacesaria, M. di Serio, R. Tesser, M. Tortorelli, R. Turco, V. Russo, A simple device to test biodiesel process intensification. Chem. Eng. Process.: Process Intensification **50**(10) (2011).
3. V. Hessel, H. Löwe, F. Schönfeld, Micromixers—a review on passive and active mixing principles. Chem. Eng. Sci. **60**(8–9 SPEC. ISS.) (2005).

4. R. Gupta, D. Fletcher, B. Haynes, Taylor flow in microchannels: a review of experimental and computational work. J. Comput. Multiphase Flows **2**(1) (2010).
5. T. Xie, L. Zhang, N. Xu, Biodiesel synthesis in microreactors. Green Process. Synth. **1**(1) (2012).
6. G. Guan, K. Kusakabe, K. Moriyama, N. Sakurai, Transesterification of sunflower oil with methanol in a microtube reactor. Industr. Eng. Chem. Res. **48**(3) (2009).

Development of Reforming Catalyst for Hydrogen Production and Its Suitability for Proton Exchange Membrane Fuel Cell

Punampriya Borgohain, Pankaj Tiwari, and Rajesh Kumar Upadhyay

Abstract Hydrogen as an energy carrier has a potential to replace fossil fuel and meet the rising demand without compromising with the environment. The high efficiency to generate power from hydrogen comes when it is integrated with proton exchange membrane fuel cell (PEMFC). However, PEMFC needs high-purity hydrogen (purity more than 99.99%) with CO content lower than 10 mg/L. Methanol steam reforming reaction has the potential to produce H_2, and its performance is mainly relying on the type of reforming catalyst which should bear the ability to have low CO selectivity. In our current work, a series of copper-based reforming catalysts were synthesized by incipient impregnation method and examined for steam reforming of methanol reaction with steam to carbon molar ratio of 3:1 and up to a temperature range of (200–350 °C) at atmospheric pressure conditions. The characterization of the prepared catalysts was carried out by using XRD diffraction, Brunauer–Emmett–Teller (BET) surface area and FESEM-EDX techniques. The promotional effect of gallium over the copper catalyst was investigated for hydrogen generation. Among all the synthesized catalysts, gallium oxide modified Cu-Fe catalyst which shows low CO selectivity (less than 1%) even at elevated temperature and higher methanol conversion compared to the undoped gallium catalyst.

Keywords Hydrogen generation · Bimetallic catalyst · Methanol steam reforming · Gallium based catalyst

1 Introduction

Fuel cell-driven power plants and electrical vehicles utilizing hydrogen (H_2) as an alternative to the existing fossil fuel are now being generated for the safeguard of our

P. Borgohain · P. Tiwari
Department of Chemical Engineering, Indian Institute of Technology Guwahati, Guwahati, Assam, India

R. K. Upadhyay (✉)
Department of Chemical Engineering and Technology, Indian Institute of Technology (BHU) Varanasi, Varanasi, Uttar Pradesh, India
e-mail: rku.che@itbhu.ac.in

V. S. Moholkar et al. (eds.), *Sustainable Energy Generation and Storage*,
https://doi.org/10.1007/978-981-99-2088-4_4

environment and also to promote sustainable energy development [1]. Hydrogen acts as a power transporter for energy purposes and a raw material for the chemical sector as a basic building element in future energy scenarios [1]. Methanol steam reforming seems to be an active area of research for the generation of hydrogen. Methanol is a form of alcohol consisting of only one carbon atom. It is a zero-emission fuel, which can be easily stored and transported. Also, methanol has a greater ratio of H/C and a low reforming temperature than other alkanes and oxygenates, viz. methane [2], glycerol [3] and ethanol [4]. In general, the methanol steam reforming reaction is indicated by Eq. 1, but it comes with the cost of two side reactions, namely water gas shift reaction (Eq. 2) and methanol decomposition (Eq. 3) [5] as given below:

$$CH_3OH + H_2O \rightleftharpoons CO_2 + 3H_2 \ \Delta H^\circ_{298K} = +49.7\,\text{kJ/mol} \quad (1)$$

$$CH_3OH \rightleftharpoons CO + 2H_2 \ \Delta H^\circ_{298K} = -41.2\,\text{kJ/mol} \quad (2)$$

$$CO + H_2O \rightleftharpoons CO_2 + H_2 \ \Delta H^\circ_{298K} = +90.7\,\text{kJ/mol} \quad (3)$$

CO formation (Eq. 2) as a by-product of reforming is one of the minor impediments to the MSR process, as the CO concentrations more than 10 ppm induce catalyst poisoning of the anodic section of low-temperature fuel cells [6]. As a result, it is important to break through this barrier in the methanol reforming process by boosting catalyst performance. The designing of the MSR catalyst must have the unique properties including high catalytic activity, low CO production and high selectivity for hydrogen generation and high stability. Generally, copper-assisted catalysts facilitate methanol steam reforming for H_2 generation and utilization owing to its improved catalytic activity. However, thermal agglomeration phenomena (sintering) of copper metal are one of the major drawbacks of copper catalyst. In the literature, various approaches for improving the performance of copper catalysts have been published [7, 8]. Some studies focus on the utilization of various catalyst supports and promoters, while others examine the impact of catalyst synthesis methods and also consider the impact of surface area of the support, high metal distribution and metal size reduction on catalytic performance. Alumina is widely used as a support due to the fact that it provides high surface area with excellent thermal stability and proper metal dispersion [9]. It has been seen that the copper-based catalyst and the selectivity of product mostly depend on the status of Cu and its dissipation over the support. To obtain this, incorporation of ZnO and ZrO_2 to copper was carried out, zinc allows higher copper dispersion, and zirconia minimizes the carbon monoxide yield [10, 11]. Apart from Cu, Fe was used as a secondary metal due to its variable oxidation states, and it has oxidation and reduction property in it. Incorporation of Fe into the catalyst increases the catalytic activity and also improves the sintering of copper [12]. One of the least studied promoters is the gallium oxide in reforming catalyst for MSR. Gallium-assisted copper zinc catalyst was prepared by coprecipitation method where introduction of Ga^{3+} ions into the Cu–Zn oxide increases the dispersion of Cu and thereby enhanced catalytic activity was obtained [13]. Finally, we have come up with a new catalyst composition where gallium is incorporated to Cu–Fe catalyst. To the best of our knowledge, no such catalyst has been designed for steam reforming of methanol process. Hence, a new class of reforming catalyst has been designed in our current work by the adding of gallium as a promoter over Cu–Fe catalyst for the generation of hydrogen with minimal amount of CO at elevated temperature.

In the current study, MSR experiments were being conducted for gallium-promoted Cu/Fe catalyst to check the enhanced activity of the catalyst for the generation of H_2. Characterization of the synthesized catalysts and testing of catalyst activity for MSR was performed, and comparison study was done between the gallium-doped and undoped catalyst.

2 Experimental

2.1 *Materials*

All the metal precursors, viz. copper nitrate trihydrate was purchased from Merck reagent grade, ferric nitrate nonhydrate from Loba Chemie, zirconyl nitrate from Loba Chemie, zinc nitrate hexahydrate from Merck, gallium nitrate hydrate and aluminium oxide from Alfa Aesar. The feedstock for the methanol steam reforming are such as methanol which was purchased from SD Fine Chemicals limited, and water used was distilled water.

2.2 *Catalyst Preparation*

All the catalysts having the composition Cu/Al_2O_3–Zn–Zr (10/90 wt%), Cu–Ga/Al_2O_3–Zn–Zr (7.5/2.5/90 wt%) and Cu–Fe–Ga/Al_2O_3–Zn–Zr (7.5/1.25/1.25/90 wt%) were prepared via incipient wetness impregnation. All the catalyst has metal loading of 10 wt% on a fixed support composition of Al_2O_3–Zn–Zr (AZZ) (73–15–12 wt%). The support composition (AZZ) was taken from earlier work from our laboratory [14]. Aluminium oxide was first crushed and sieved to the size of 200 μm. Sequential doping of zinc nitrate followed by zirconyl nitrate on alumina was done via incipient impregnation method. Then 10 wt% metal loading was carried out to 90 wt% of the support by incipient wetness impregnation method. Once the metal loading was done, the prepared catalyst was allowed to dry and then calcined at 500 °C for 6 h.

2.3 *Catalyst Characterization*

The BET surface areas of the synthesized catalysts were achieved using N_2 adsorption–desorption data taken at liquid N_2 (− 196 °C) temperature on a Micromeritics ASAP 2100 equipment. The samples were degassed at 200 °C for 3 h in vacuum condition before the analysis to remove any kind of impurities present in it.

The X-ray diffraction patterns of all the synthesized catalysts were analysed utilizing Smart lab X-ray diffractometer functioning at 45 kV and 112 mV with Cu–Kα radiation source at wavelength = 1.544 Å. The scanning range $2\theta = 10–80°$ with a scanning speed of 4°/min in order to analyse the structure.

The morphological structure of the prepared samples after pre-treatment was analysed by Zeiss Sigma 300 Field Emission Scanning Electron Microscope (FESEM).

3 Result and Discussion

3.1 *Catalyst Activity Testing*

Experiments on catalytic activity were conducted in an electric furnace utilizing a constant flow tubular fixed-bed reactor having an inner diameter of 9 mm. All the tests were carried out at an operating temperature of 200–350 °C with 500 mg of catalyst kept in the tubular reactor. The catalyst reduction was performed in situ with the help of premixed H_2/N_2 (10/90 (v/v)) gas flow at a rate of 40 ml/min with subsequent heating at 400 °C for 3 h. The reaction for methanol steam reforming using a 1:3 molar mixture of CH_3OH and H_2O was carried out inside the tubular reactor. The premixed methanol and water were then delivered to the vaporizer, which was kept at a temperature of around 180 °C. The vaporized feed was injected into the reactor with a stream of N_2 gas at a flow rate of 30 ml/min. The steam reforming process was then started at the specified reaction temperature. The analysis of the obtained products was conducted using a gas chromatography system consisting of a thermal conductivity detector.

3.2 *Characterization of Catalysts*

Figure 1 shows the X-ray diffraction (XRD) patterns of the catalyst displayed a diffraction pattern at 2θ of 35.54° for Fe_2O_3 structure in Cu–Fe–Ga–AZZ catalyst. In monometallic Cu–AZZ catalyst, a distinct peak of CuO at 2θ of 35.56°, 38.5° was observed. No characteristic peak of ZnO, ZrO_2 and Al_2O_3 was found; the reason behind this is due to the amorphous state and their well-dispersed nature. The absence of Ga peaks might be due to high dispersion and very fine particle size [15].

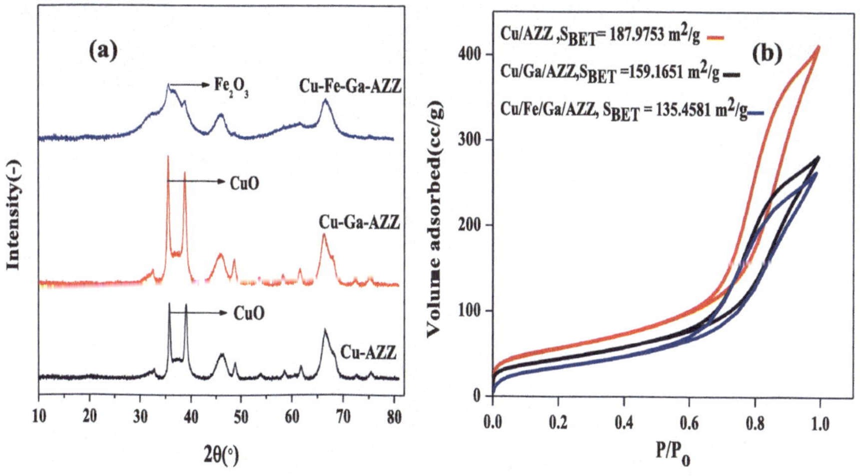

Fig. 1 XRD patterns and BET surface area of the prepared catalysts

Figure 1b displays the BET surface area of all the prepared catalyst. N_2 adsorption–desorption isotherms of all the prepared catalysts were acquired at liquid nitrogen temperature (− 196 °C) on a Micrometrics ASAP 2020M apparatus. The catalysts were degassed at 200 °C for 3 h and were observed to follow Type IV isotherm defining the characteristic of mesoporous materials [16].

Figure 2a–c shows the FESEM images of all the prepared catalyst. It can be seen that the incorporation of gallium the morphology of the catalyst changed to loose needle bar-shaped structure [15].

The MSR activity of all the catalyst with different loading of metal on Al_2O_3–Zn–Zr (AZZ) for methanol conversion as a function of temperature was represented in Fig. 3a–c. It has been seen that the activity of the bimetallic catalyst Cu–Fe–Ga/AZZ was highest in terms of methanol conversion among the other two prepared catalysts. The methanol conversion increases with the incorporation of the secondary metal Fe due to the fact that Cu and Fe developed a strong synergistic effect which provided a better performance in the catalyst [17]. Moreover, MSR is an endothermic reaction, as such methanol conversion increases with an increase in temperature.

Overall comparison of the prepared catalysts can be seen from Fig. 3d that at 200 °C and 250 °C gallium-based catalyst showed zero CO selectivity which is not seen in monometallic Cu/AZZ catalyst. The reason behind may be due to the fact that introduction of gallium into the Cu–Fe oxide originates a nonstoichiometric geometry containing interstitial Cu^+ ions, resulting in very fine and well-dispersed copper particles [18].

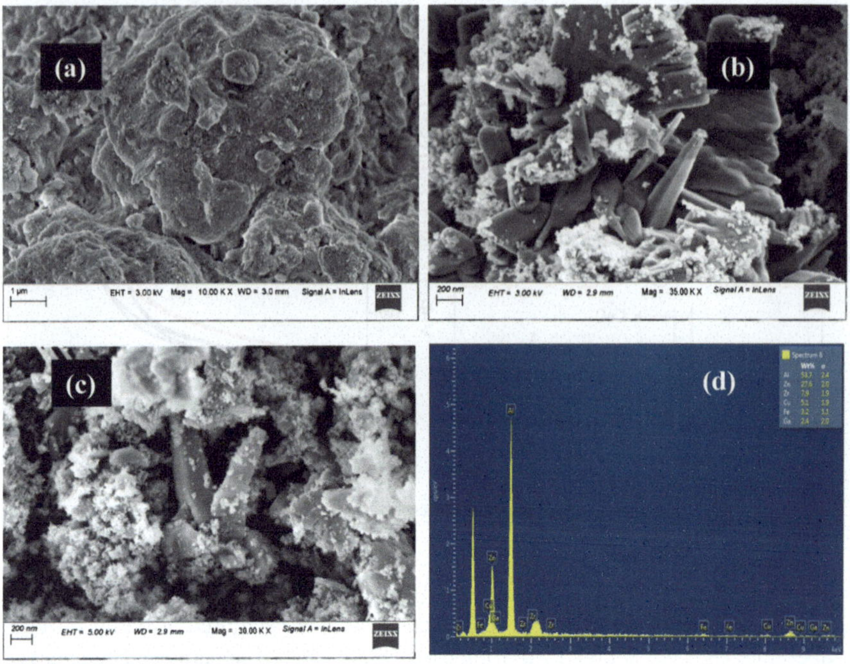

Fig. 2 FESEM images of all the prepared catalysts **a** Cu/AZZ, **b** Cu/Ga/AZZ, **c** Cu/Fe/Ga/AZZ and **d** EDX of Cu/Fe/GA/AZZ catalyst

4 Conclusion

The current work investigated the catalytic activity of copper-based catalyst supported on alumina zinc zirconia support. Catalysts with Cu/Al_2O_3–Zn–Zr (10/90 wt%), Cu–Ga/Al_2O_3–Zn–Zr (7.5/2.5/90 wt%) and Cu–Fe–Ga/Al_2O_3–Zn–Zr (7.5/1.25/1.25/90 wt%) were prepared utilizing impregnation method. Both monometallic and bimetallic catalysts were synthesized, and the catalytic activity for methanol steam reforming was examined. Catalyst characterization was done by FESEM, EDX, XRD and BET. Gallium-modified Fe–Cu catalyst reported the highest catalytic activity with regard to methanol conversion and low selectivity of CO. Moreover, the interaction between CuO and Fe_2O_3 adds promotional effect such as high dispersion of copper particles and reduction of copper in Cu–Fe–Ga/AZZ catalyst. With the addition of gallium as promoter, zero CO selectivity was observed at 200 and 250 °C. Therefore, it was inferred that an improvement in the catalytic activity of Cu–Fe–Ga/Al_2O_3–Zn–Zr catalyst could prove to be a potential catalyst for steam reforming of methanol and suitable for PEMFCs applications.

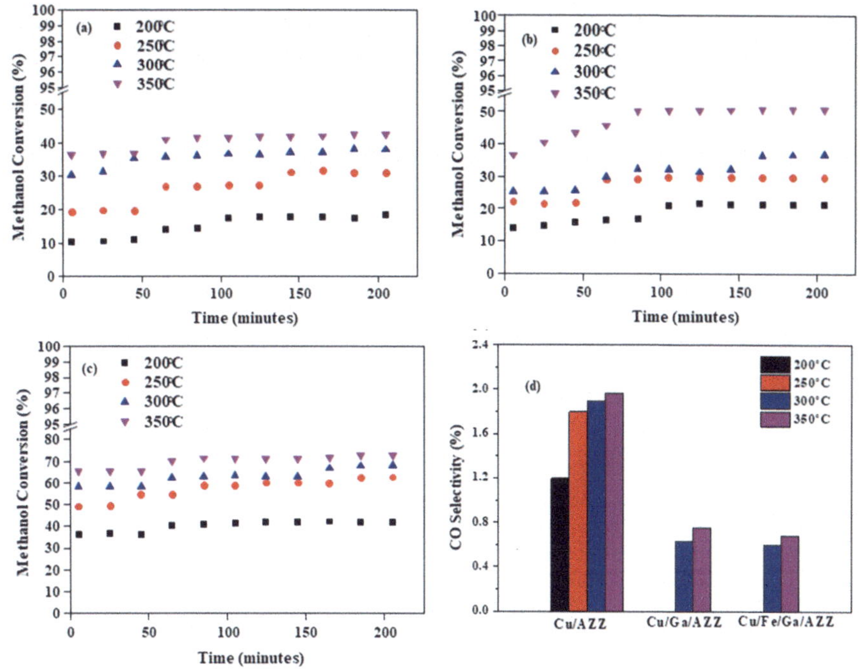

Fig. 3 Catalyst activity testing **a** Cu/AZZ, **b** Cu/Ga/AZZ and **c** Cu/Fe/Ga/AZZ catalysts in terms of methanol conversion and **d** overall comparison on CO selectivity of all the catalysts

Acknowledgements The funding of our research work was given by the Department of Science and Technology, India, under the agreement number TMD/CERI/MDME/2016/81, and the authors would also like to acknowledge Central Instrument Facility at IIT Guwahati for providing me the instruments for analysis.

References

1. G.A. Olah, Towards oil independence through renewable methanol chemistry. Angewandte Chemie—Int. Ed. **52**(1), 104–107 (2013)
2. D.R. Palo, R.A. Dagle, J.D. Holladay, Methanol steam reforming for hydrogen production. Chem. Rev. **107**(10), 3992–4021 (2007)
3. R.O. Idem, N.N. Bakhshi, Production of hydrogen from methanol over promoted coprecipitated Cu-Al catalysts: the effects of various promoters and catalyst activation methods. Ind. Eng. Chem. Res. **34**(5), 1548–1557 (1995)

4. H. Song, Y. Liu, H. Bian, M. Shen, X. Lin, Energy, environment, and economic analyses on a novel hydrogen production method by electrified steam methane reforming with renewable energy accommodation. Energy Convers. Manage. **258**(March), 115513 (2022)
5. S.B. Bagherzadeh, M. Haghighi, N. Rahemi, Novel oxalate gel coprecipitation synthesis of ZrO_2-CeO_2-promoted CuO-ZnO-Al_2O_3 nanocatalyst for fuel cell-grade hydrogen production from methanol: influence of ceria-zirconia loading. Energy Convers. Manage. **134**, 88–102 (2017)
6. R. Koch, E. Lopez, N.J. Divins, M. Allue, A. Jossen, J. Riera, J. Llorca, Ethanol catalytic membrane reformer for direct PEM FC feeding. Int. J. Hydr. Energy **38**(14), 5605–5615 (2013)
7. S. Sá, H. Silva, L. Brandão, J.M. Sousa, A. Mendes, Catalysts for methanol steam reforming—a review. Appl. Catal. B **99**(1–2), 43–57 (2010)
8. S.T. Yong, C.W. Ooi, S.P. Chai, X.S. Wu, Review of methanol reforming-Cu-based catalysts, surface reaction mechanisms, and reaction schemes. Int. J. Hydrogen Energy **38**(22), 9541–9552 (2013)
9. R.L. Manfro, N.F.P. Ribeiro, M.M.V.M. Souza, Production of hydrogen from steam reforming of glycerol using nickel catalysts supported on Al_2O_3, CeO_2 and ZrO_2. Catal. Sustain. Energy **1**, 60–70 (2013)
10. H. Silva, C. Mateos-Pedrero, P. Ribeirinha, M. Boaventura, A. Mendes, Low-temperature methanol steam reforming kinetics over a novel CuZrDyAl catalyst. React. Kinet. Mech. Catal. **115**(1), 321–339 (2015)
11. C. Mateos-Pedrero, H. Silva, D.A.P. Tanaka, S. Ligouri, A. Iulianelli, A. Basile, A. Mendes, CuO/ZnO catalysts for methanol steam reforming: the role of the support polarity ratio and surface area. Appl. Catal. B Environ. **174–175**, 67–76 (2015)
12. C.L. Yu, S. Sakthinathan, B.Y. Hwang, S.Y. Lin, T.W. Chiu, B.S. Yu, Y.J. Fan, C. Chuang, $CuFeO_2$–CeO_2 nanopowder catalyst prepared by self-combustion glycine nitrate process and applied for hydrogen production from methanol steam reforming. Int. J. Hyd. Energy (2020)
13. W. Tong, A. West, K. Cheung, K.M. Yu, S.C.E. Tsang, Dramatic effects of gallium promotion on methanol steam reforming Cu-ZnO catalyst for hydrogen production: formation of 5Å copper clusters from Cu-$ZnGaO_x$. ACS Catal. **3**(6), 1231–1244 (2013)
14. R. Sharma, A. Kumar, R.K. Upadhyay, Bimetallic Fe-promoted catalyst for CO-free hydrogen production in high temperature-methanol steam reforming. ChemCatChem **11**(18), 4568–4580 (2019)
15. C. Rameshan, H. Lorenz, M. Armbuster, I. Kasatkin, B. Klotzer, Impregnated and co-precipitated Pd–Ga_2O_3, Pd–In_2O_3 and Pd–Ga_2O_3–In_2O_3 catalysts: influence of the microstructure on the CO_2 selectivity in methanol steam reforming. Catal. Lett. **148**(10), 3062–3071 (2018)
16. A. Pohar, S. Hočevar, B. Likozar, J. Levec, Synthesis and characterization of gallium-promoted copper-ceria catalyst and its application for methanol steam reforming in a packed bed reactor. Catal. Today **256**(P2), 358–364 (2015)
17. L. Cao, M. Lu, G. Li, S. Zhang, Hydrogen production from methanol steam reforming catalyzed by Fe modified Cu supported on attapulgite clay. React. Kinet. Mech. Catal. **126**(1), 137–152 (2019)
18. M. Lachowska, Steam reforming of methanol over Cu/Zn/Zr/Ga catalyst: effect of the reduction conditions on the catalytic performance. React. Kinet. Mech. Catal. **101**(1), 85–91 (2010)

Mechanistic Aspects of Enhanced Kinetics in Sonoenzymatic Processes Using Three Simultaneous Approaches

Karan Kumar and Vijayanand S. Moholkar

Abstract Enzyme-mediated processes are being popularized in industries due to its cost-effectiveness, low energy intake, nontoxic, and eco-friendly nature. Numerous studies have reported the enhancement of the reaction kinetics of enzyme-catalyzed processes with aid of advanced techniques such as microwaves, supercritical fluid, ionic liquids, ultrasound, and chemical modifiers. However, very little is known about the mechanism of enhanced reaction kinetics. Herein, we propose a methodology that includes amalgamation of three simultaneous approaches (computational and experimental), viz. quantum chemical simulation, mathematical modeling, and secondary structural analysis of enzymes to unravel the mechanistic aspect of enhanced reaction kinetics with the application of sonication.

Keywords Ultrasound · Molecular docking · Density functional theory · Kinetic mechanism · Secondary structure

1 Introduction

In recent decades, scientific interests in synthesis of platform chemicals and essential commodities are being immensely shifted to become sustainable and environmental friendly; therefore, the presence of green chemistry in most chemical industries is inevitable [1–4]. These interests are supported by the rapid growth in demand for products with growing complexity. In this context, enzyme-based technologies are emerging with application as industrial catalysts [5–7]. Enzymes are highly efficient biocatalysts made up of large bio-macromolecules composed of amino acids [8, 9]. Enzyme-mediated processes are winning the competition against other catalytic

K. Kumar (✉) · V. S. Moholkar
School of Energy Science and Engineering, Indian Institute of Technology Guwahati, Guwahati, Assam 781039, India
e-mail: karan18@iitg.ac.in

V. S. Moholkar
Department of Chemical Engineering, Indian Institute of Technology Guwahati, Guwahati, Assam 781039, India

V. S. Moholkar et al. (eds.), *Sustainable Energy Generation and Storage*,
https://doi.org/10.1007/978-981-99-2088-4_5

processes due to their cost-effectiveness, low energy requirements, nontoxic, and eco-friendly nature [7, 10]. Numerous studies have reported the enhancement of the reaction kinetics of enzyme-catalyzed processes with aid of advanced techniques such as microwaves [11], supercritical fluid [12], ionic liquids [13], ultrasound [9, 10, 14], and chemical modifiers [15]. The native (or wild-type) enzymes show optimal enzymatic activities in their optimal temperature range (mostly from 30 to 50 °C) [16]. In this temperature range, various chemical processes have slow reaction rates due to which the chemical process consumes lot of time, therefore, less attractive on economic grounds [9]. Most of the chemical reactions in chemical industries are carried out at higher temperature range that may denature the enzymes [17]. Hence, this major bottleneck of thermal stability of enzymes is a critical aspect in enzyme engineering for increasing the optimal temperature range of enzymes [18]. Thermal stability of enzyme at higher temperatures is an important feature that contributes to the industrial applications of any enzyme. Other common drawbacks with industrial application of enzymes include the high cost of the enzymes, their inhibition or deactivation by excess substrates/solvent molecule, and low enzyme recovery after completion of reaction [19].

Recent developments in various scientific domains have opened up multiple ways of improving the aforementioned limitation of enzyme application in industries [20]. Microbiological techniques (such as metagenomics) and genetic techniques (e.g., directed evolution, site-directed mutagenesis, etc.) have helped in higher enzyme production with desired enzymatic properties as compared to the wild-type enzymes [16]. Initially when invented, immobilization of enzymes on solid support was just a technique to resolve the issue of its solubility in reaction systems, but in present days, this has become an important and powerful tool for improving various other desirable industrial features which include thermostability, enzymatic activity, substrate selectivity and specificity, enhanced purity, reducing inhibition, and improving resistance to chemicals [21–23]. Although the aforementioned tools such as genetic and microbiological are very attractive and capable to resolve various issues associated with enzyme catalysis, these methods are much expensive and require ample amount of time to make the enzymes desirable for industrial-scale application [9, 24]. Thus, it is the need of the hour to identify the sustainable and time saver techniques such as ultrasonic irradiation (or sonication) that intensify the enzyme-catalyzed processes without effecting enzymatic structure. Sonication is a sustainable, eco-friendly, and less time-consuming approach as compared to its alternatives. Sonication has been shown to intensify the reaction rate and thus overcome the issues related to requirement of long reaction time in conventional enzymatic processes [14, 24–27]. Sonoenzymatic methods offer a lucrative approach to intensify the chemical and biophysical processes based on the cavitational effects of ultrasound that is introduced in the liquid systems. Application of ultrasonic reactors helps in improving the reactions by improving reaction kinetics that are having intrinsically slow rates, lower product yields, and need expensive catalysts for progress, thus making them cost-effective [9, 28, 29].

1.1 Aim and Scope

However, numerous studies are available that report the enhancement of the reaction kinetics of chemical reaction by using ultrasound [9, 10, 14], very little is known regarding how ultrasound irradiation helps in enhancing the reaction kinetics [9]. Herein, we propose a methodology which is amalgamation of three simultaneous approaches (including both computational as well as experimental techniques), viz. quantum chemical simulation, mathematical modeling, and secondary structural analysis of enzymes to unravel the mechanistic aspect of enhanced reaction kinetics with the application of sonication. This amalgamated approach can be represented as vertices of a solution triangle that include (1) estimation of secondary structural changes in enzymes, (2) computational techniques such as molecular docking, and (3) mathematical modeling to understand and explain kinetics of the reaction. It is known from the previous literature that ultrasound irradiation has impacts on the secondary structure composition of enzymes which can be estimated via techniques such as deconvolution of FT-IR spectra, CD spectroscopy, and Fluorescence spectroscopy. To explain the impacts of structural changes on reaction kinetics, one can perform molecular docking analysis that reveals the molecular mechanism of enzymatic reactions and product formation. Furthermore, mechanistic investigation by using mathematical models can elucidate the thermodynamic and macroscopic effects of ultrasound on reaction kinetics to make it faster. As a proof of concept, we have recently published a research article [9, 27], in that, we have reported the mechanistic accounts of sonoenzymatic synthesis of n-butyl levulinate. The aim of this article is to provide the in-depth mechanistic insights into the various effects of ultrasound on kinetic parameters of enzyme-catalyzed process.

2 Salient Features of Enzymes and Enzyme-Catalyzed Processes

2.1 Enzymes as Biocatalysts in Industries

As described in the introductory section, enzymes are protein molecules made up of long chains of amino acids that act as macromolecular protein catalysts (or biocatalyst) [7, 30]. The long chains are linear condensed polymers joined through amide linkage called as polypeptide chains [31]. These polypeptide chains fold in precise three-dimensional structure (also called as tertiary structure) that make functional form of a protein. Some enzymes are in their quaternary structure conformation which is made up of two or more 3D structural forms consisting multiple polypeptide chains (also known as globular proteins) [32]. Enzymes catalyze all non-spontaneous chemical reactions in the living cells or organisms [5, 17]. As biocatalysts, they are very sensitive to physiological conditions, such as temperature and pH, and can only catalyze the reaction having their specific reactants, unlike their alternatives such as inorganic catalysts. Similar to metallic or non-metallic catalysts, enzymes accelerate the reaction rate of any given chemical reactions by lowering the required activation energy with a higher turnover rate. In industry, enzymes are produced using submerged fermentation which involves the cultivating and growing of microorganism in very contained and closed environment [33–35]. During their life cycle, microorganisms break down the nutrients that are provided in the culture medium and produce desired enzymes. Mostly, these enzymes are produced extracellularly and secreted in medium. Enzymes offer many advantages that include (a) cost-effectiveness (low cost raw materials, limited or no use of costlier chiral resolving agents, less labor and equipment cost, board range of substrate specificity, etc.), (b) higher productivity (minimum synthesis steps, larger capacity, greater yields, high regio- and enantioselectivity under mild conditions), (c) enhanced quality of intermediates (less impurity or lower by-products, high chemo-, regio-, stereo-selectivity), and (d) more greener and environmental route (minimal waste production and solvent usage, less energy intensive) [6, 7, 36].

Last few decades have witnessed numerous studies reported by authors working in the area of ultrasound-assisted sonoenzymatic processes [9, 10, 13, 24, 25, 29] using enzymes as biocatalyst with plethora of data is available for both positive and negative impacts on the chemical reaction systems. Enzymes of microbial origin are more commercially important as they may originate from bacteria or fungi. Moreover, microbial enzymes have low cost of production, substantial stability, and extensively available than enzymes derived from plant and animal sources. An enzyme family, such as lipases, occurs naturally in living organisms that catalyze and regulate biochemical processes [23]. Besides, they have atypical mechanism of action called "interfacial activation", which is described later in this article [20, 37, 38]. Despite lipases derived from natural sources have many advantages over other catalyst used in organic chemistry, they typically lack the desirable features that are suitable for industrial-scale reactions which were discussed earlier in this section [20]. Figure 1 represents the action of enzyme on a glyceride substrate. As a biocatalyst enzyme can act at the different nucleophilic sites of the substrate resulting in various types of substitution mechanism in the reaction systems.

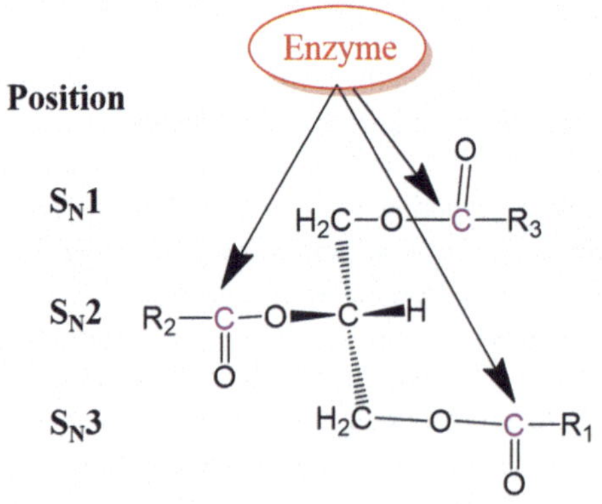

Fig. 1 Schematic representation of enzyme acting as biocatalysts by catalyzing various types of nucleophilic substitution reaction mechanisms: S_N1, S_N2, and S_N3 based on the substrate's concentration

2.2 Importance of Structural Features and Catalytic Sites of Enzymes

Among the most special features of enzymes is their specificity in the activation and inhibition of catalytic domains. This specificity toward substrate or inhibitor binding to the enzyme's active site comes from their unique tertiary (or three-dimensional) structure. The stability of enzyme's tertiary structure comes from intramolecular interactions such as non-bonded, coulombic, and hydrogen bonds between the monomeric amino acids present in the polypeptide chains. The 3-dimensional structure or geometric form of functional enzymes form depends on the type of amino acids present. The catalytic cavity or active site is built up of only few amino acids. Catalytic cavity or binding site is the only small center in the functional structure of enzymes where interaction of enzyme–substrate and enzyme-product transpires. The type of amino acids available and the ultimate shape of catalytic cavity determine which substrate will have access to the binding site of the enzyme. Thereby, it is being ensured that only substrates that are complementary to the binding pocket (similar to specific key for a lock) have accessibility and initiate the catalytic activities of enzymes. This specificity in binding can be a type of either absolute specific or functional group-specific. In enzymatic reactions, binding of the substrate to the enzyme's active site is very crucial step. In addition to substrates, an enzyme may bind specifically to the inhibitors, activators, allosteric activators, etc., at or near or at the regions distant from the active site pocket. Therefore, having the knowledge of functional structure of enzyme is essential for studying any enzyme-catalysis reaction. Computer simulation has proven to be a powerful tool in this regard to solve protein structure as well as to understand enzyme–substrate–solvent interactions. Many of the enzymes have been solved using experimental methods (e.g., NMR, X-ray crystallography, cryo-electron microscopy) and computer simulation methods (homology modeling and comparative modeling against template structures).

2.3 Challenges in Enzyme-Catalyzed Processes

With the emergence of enzyme for synthesizing innumerable platform chemicals and essential commodities, the major disadvantages in enzymes catalyzed processes are slow reaction rates, expensiveness, and solubility in reaction systems as compared to chemical processes. However, the techniques such as immobilization can resolve the issues of high cost and reusability, the immobilization makes the system heterogeneous. The heterogeneous systems require additional energy to overcome the resistance in mass transfer that results in lowering down the reaction rate. Additionally, the lower reaction rate may occur because of the product accumulation on the enzyme surface (known as product inhibition). Therefore, it is crucial to intensify these processes using sustainable techniques to improve the reaction rate from an industrial perspective and for greater applicability of enzymes in the international market. A few important industrial enzymes, such as lipase, face the challenges of interfacial activation. Glycerides are the natural substrates of lipases. Glycerides suffer with lower solubility in aqueous solvents due to which they form droplets that are insoluble in aqueous phase, where enzymes must act. Lipases are also called "interfacial enzymes" because they have an atypical mechanism of action, known as interfacial activation, due to which lipases can act at the aqueous-apolar interphase by being adsorbed on the hydrophobic surface of the glyceride droplets. This mechanism occurs due to the presence of a hydrophobic lid covering the catalytic cavity (or active site) of lipases. The hydrophobic lid also consists an internal hydrophobic pocket that can interact with the hydrophobic amino acids of the catalytic cavity and an external hydrophilic domain that interacts with the reaction medium. When the hydrophobic lid moves, it exposes a huge hydrophobic domain (pocket) consisting the catalytic cavity which becomes accessible to the reaction medium. This action results in the "open" and active form of the lipase. An enzyme having this large hydrophobic pocket results in extremely unstable conformation which becomes challenging when present in water (or aqueous) media.

3 Ultrasound: Theory and Its Applications in Enzyme Catalysis

Ultrasound is an acoustic (or sound) wave, typically in the frequency range of 20 kHz to 5 MHz (that are beyond the human audibility), and requires a medium to propagate. Ultrasonic waves propagate into liquid medium in the form of a series of alternating compression and rarefaction cycles. This process creates a phenomenon in propagating medium which is commonly known as cavitation (or sonication). Acoustic waves disseminating in liquid medium at frequency in the ultrasonic range have the length of their wavelengths in multiple magnitude of interatomic bond length or molecular dimensions. The different stages of cavitation include generation of cavities (which are nothing but microbubbles), followed by growth of microbubbles to its maximum size, and finally violent collapse that leads to dissipation of significant energy locally. This considerable energy release during the collapse of microbubbles generates outward propagating shear forces and shockwaves that results in intense turbulence in the local environment. The resultant mechanical effects expedite the reactant blending and also intensify the heat and mass transfer during the process. Additionally, ultrasonic irradiation also introduces the chemical effects in the reaction system. The collapsing microbubbles can catalyze the splitting of water along with dissolved oxygen molecules in water. This phenomenon can also create highly reactive radical $^{\cdot}OH$ and radical $^{\cdot}OOH$ free radicals. In due course, these reactive species can react or recombine with other molecules available in reaction system. Thus, sonication can chemically assist the progress of reaction in the local hot spots by the radical mechanisms or physically by accelerating the mass transfer rates in the reaction system. The sudden pressure gradient caused by cavitation drags the liquid during the cycle of negative pressure around the site of bubble collapse (known as weak spot or sites of nucleation) that carry few gaseous impurities and forms a void. The necessary and essential characteristics of cavitation process to bring chemical changes in the reaction system is positive difference in the intensity of acoustic waves supplied and cavitation threshold of the reaction systems that depends on the effective volatility of reactant to enter into the microbubble generated during cavitation. Another important requirement is availability of stable and transient cavitation for the chemical and physical changes to occur in the reaction systems. There have been two theories proposed to describe the impacts of cavitation on the reaction system, viz hot spot theory and the electrical theory. The most accepted and recognized theory for explaining the impacts of cavitation in reaction system is the first one among two.

(a) The hot spot theory: This theory assumes local hot spot are formed during the sudden bubble collapse that escalates the temperature and pressure as high as 5000 K and 500 bar, respectively.
(b) The electrical theory: This theory assumes that an immense electric field gradient is formed because of the electric charge generation on the cavitation bubble's surface, that can break the chemical bonds during its bubble implosion.

3.1 Applications of Sonication in Enzyme-Catalyzed Processes

The sonochemistry is the study of understanding the effects of ultrasonic waves in liquid media in response to the cavitation generated by ultrasound. The first study on sonochemistry was reported in the early twentieth century which was renaissance in the decade of 1980s. It was seen that the ultrasonic waves could create a phenomenon called cavitation in the liquid systems that creates the gradient of temperature and pressure in the local microenvironment of the reaction media. Therefore, this field of research attracted the attention of the researchers working in the biotechnology and allied area. The interesting features of sonication-assisted process that include even at lower ultrasonic frequencies can enhance the solubility of the substrates and catalysts in the liquid system thereby enhancing the mass transfer and maintaining the homogeneity of reaction systems. Apart from the aforementioned physical effects that sonication brings in the reaction systems, it is also known to have impacts on the weak intermolecular/interatomic interactions resulting in conformational changes. The propagation mechanism of ultrasonic waves in liquid systems is discussed in Sect. 3. The propagation of ultrasonic waves initiates mechanical vibration, also known as cavitation, in the liquid systems. During the cavitation of liquid systems, microbubbles are formed that develop and finally collapse. These microbubbles that are oscillatory in nature are the hub for accumulation of ultrasonic energy during the development and growth phase, therefore, due to the sudden collapse results in almost instantaneous escalation in local pressure (approx. 1000 bar) and temperature and energy (approx. 5000 K) in short time-span. As discussed in Sect. 3, ultrasound can make changes in the chemical reactions that occur in homogeneous reaction systems through generation of free radical. However, ultrasound cannot bring significant modification in ionic reactions that are involved in reaction pathway or reactive intermediates. Likewise, the heterogeneous systems, the reactions that progress through the means of ionic intermediates, also experience positive changes during cavitation. However, these changes may not have economic feasibility because the agitated systems can also bring such modifications. Therefore, the chemical reaction systems that progress through generation of either ionic and free-radical species in heterogeneous systems and application of ultrasonic irradiation can help in boosting the reaction rates and better formation of products, and pathway nature can also be improved. Additionally, sonication can provide better catalytic effects. The notable benefits that can be obtained with the sonochemical synthesis in close similarity with the green chemistry include (a) increased product selectivity, (b) limited use of hazardous solvents, (c) less-energy requirements, (d) reduced in the reaction time, and (e) ameliorated utilization of the catalyst and raw materials. During initial time when the application of ultrasound in biological process was invented, it was being used either for enzyme deactivation or to disrupting the microbial cells. Presently, the application of ultrasonic irradiation is not limited to enzyme deactivation or cell disruption, and it has begun to emerge as a powerful tool for the enzyme activation (this happens during mild frequency conditions). The integration of ultrasonic technique in any enzymatic reactions not only enhances product yield but also expedites the reaction rate with greater product specificity and selectivity. For obtaining the maximum yield, it is very much essential to understand the mechanistic impacts of ultrasound irradiation on reaction systems especially on the enzymes. Figure 2 schematically represents the impacts of sonication in enzyme-catalyzed reactions. When sonication is employed at mild and favorable conditions, it helps in widening the catalytic cavity of the enzymes that results in the greater accessibility of reactants to the catalytic site of enzyme. However, very intense cavitation can also denature the enzymes and make them permanently dysfunctional.

3.2 Applications of Sonication in Other Enzymatic Processes

Challenges discussed earlier in Sect. 2.3 on the issues related to cost and solubility of enzymes can be resolved through immobilization techniques. Previously, immobilization technique was used to resolve the issues of enzyme solubility in the reaction systems that made them impure and very tedious to recover from the reaction system. Nowadays, immobilization has proven to be a powerful tool that improves many desirable features of enzymes with industrial perspective such as higher stability, enhanced activity, greater selectivity and specificity, purity, reduced inhibition, or resistance to chemicals. Immobilization also permits the higher recyclability of enzymes in chemical synthesis process thus reducing the production cost. The most applied enzyme in industry, lipase, has several advantages as a catalyst that include requirement of milder reaction conditions and minimal production of side

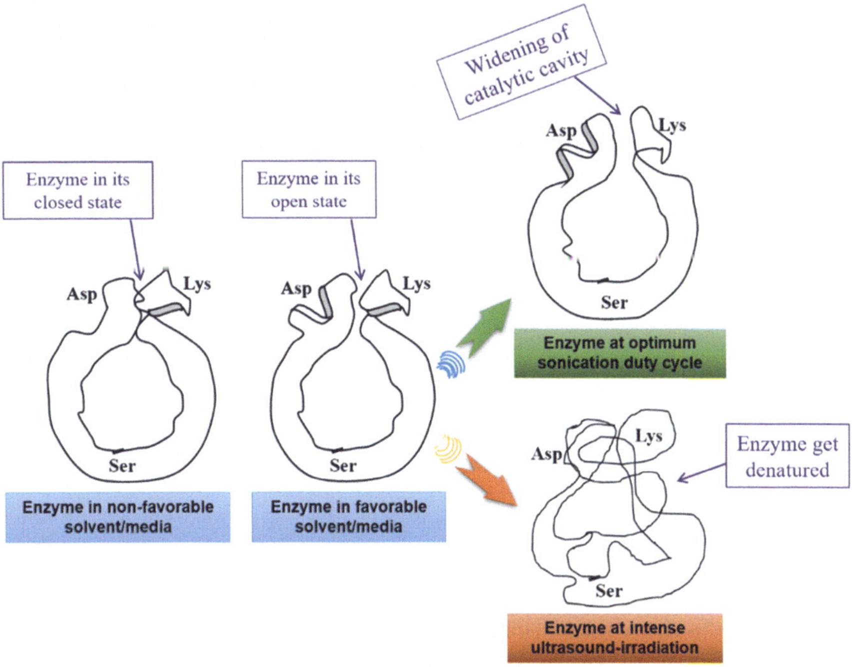

Fig. 2 Application of sonication in enzymatic reactions. Sonication helps in widening the catalytic cavity of enzyme at optimum duty cycle and denaturation at higher duty cycle

products. The product yield can be enhanced by using enzymes but the major bottleneck is that enzymes exhibit a sluggardly reaction rate. Therefore, intensifying the enzymatic process using novel approach such as ultrasound irradiation is need of the hour. As mentioned in the previous sections, enzyme immobilization is very much required for most of the enzymes used in industries, and the same goes for the applicability of lipases. Lipases have two conformations that are in equilibrium, but presence of oil drops helps the open conformation of lipase to adsorb on the oil drops that has hydrophobic surface. This activity and condition favor the equilibrium toward the open lipase conformation and permit them to attack on the glycerides (substrate of lipase). This mechanism of catalysis may create a significant problem during handling of lipase. This physical property of lipases allows them to being adsorbed on any hydrophobic surface. For example, due to this mechanism, lipases could form dimeric aggregates by interacting with any two open conformations resulting an altered property or may interact with other proteins that are hydrophobic in nature that also alter the enzyme properties. This event can cause problem during immobilization of lipase on a solid support, due to isolation of lipase from the reaction media and less probable interactions with external droplets of hydrophobic substrates with lipase, thereby decreasing the chances of interfacial activation. Familiarizing with the interfacial activation phenomenon of lipases can assist in lipase immobilization

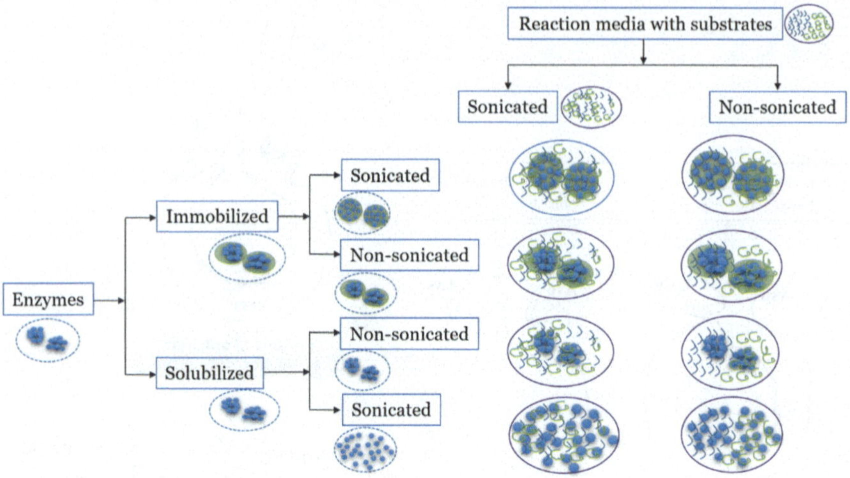

Fig. 3 Application of sonication in immobilization of enzymes on solid support and better solubilization in reaction systems

on solid support. As an example, immobilized lipases in correct orientation have been reported to utilize as a chromatography matrix that helps in purifying other lipases due to the occurrence of interaction between the open conformation of two lipases. The dimeric agglomerated lipase conformations usually exhibit altered properties and reduced catalytic activities as compared to monomeric lipase.

Many authors have reported the application of ultrasound irradiation to enhance the enzyme immobilization process and homogenizing the reaction system. The application ultrasound in such processes is summarized in the form of schematic representation given in Fig. 3. The figure represents two conditions where ultrasound is being applied. As discussed in Sect. 3, sonication assists in homogenizing the reaction systems by events such as micromixing during cavitation.

4 Mechanistic Approach to Understand Ultrasound-Assisted Enhancement of Reaction Kinetics

As discussed in Sect. 3, the ultrasound with its physical effects helps in micromixing, improving the mass transfer by effective contact of substrates with enzymes and thereby improving reaction kinetics. However, it is utmost important to understand the mechanisms by which such ultrasound operates. In the next subsections of this section, we will talk about insight into reaction kinetics through mathematical modeling, estimating the qualitative and quantitative impact of sonication on enzyme structure, and also applications of quantum mechanics/molecular mechanics

(QM/MM) and simulation techniques to understand the molecular mechanism of enzymatic reactions.

4.1 *Mathematical Modeling as a Tool to Understand the Enzyme Catalysis*

General discussion on understanding action mechanism of enzymes*:* Understanding the enzyme kinetics for a given chemical reaction system provides us the critical information regarding qualitative description of the order of substrate attachment and product release from the enzyme. This also provides us the quantitative information regarding the rate-limiting steps of any given reaction. Understanding the action mechanism of enzyme is nothing but (a) the identification of intermediates in a given reaction, (b) the identification of acid–base catalyzing group on the enzyme, (c) the role of any co-factor, and (d) the nature of transition state for the reaction catalyzed by enzyme. To deduce this information, variety of experiments/techniques need to be conducted/studied, e.g., mathematical modeling of reaction kinetics. The rate equation in the mathematical form is a function of substrate concentration that limits the kinetic mechanism of enzyme. Studying the inhibition patterns of products or dead-end inhibitors versus different substrates helps us to determine the rate-limiting step of the given chemical reaction. To fully understand the kinetic analysis of enzymatic reactions, isotope exchange and partitioning studies are required. In order to deduce the chemical mechanism, studies such as pH-dependent variation of kinetic parameters need to be conducted that identify the essential protonation states, and acid–base catalysts of the substrate-binding and enzyme catalysis. Since enzymes are nothing but proteins that are made up of long chain of amino acids, a logical approach in understanding the kinetic mechanism is to study the general features of structure of enzyme.

Available kinetic models that describe enzyme catalysis*:* For developing any kinetic model of enzymatic reaction, the action of that particular enzyme toward substrate should be known. To test and validate the kinetic model, the experimental data for enzyme concentration and reacting species are required that should be quantified with minimum error and interpreted accurately. If a chemical reaction involves two or more substrates, multiple experiments need to be carried out with considering the concentration effects of all substrates meticulously so that one can identify the action of all substrates on enzyme. The enzyme-catalyzed reactions suffer with issues such as selectivity which is drastically reduced due to the undesirable phenomenon termed as inhibition. Depending on the mode/type of inhibition, it can alter the enzymatic activity temporarily or permanently. During the inhibition, the catalytic sites consisting catalytic amino acids are inhibited either reversibly or irreversibly by (i) substrates present in the reaction systems, (ii) one or more products formed during progress of enzymatic reaction, or (iii) both substrates and products. The type of reversible inhibition could be of any of three forms, competitive, noncompetitive, or uncompetitive that depend on the type of interaction of inhibitors with enzyme. Several kinetics models have been proposed based on the number of participating substrates in the enzymatic reaction kinetics. Among the most common kinetic models employed to understand enzyme catalysis are monosubstrate reaction model (e.g., Michaelis–Menten kinetic model and Haldane kinetic model). The multi-substrate reaction model includes rapid equilibrium bi-substrate models such as ordered, random-order kinetic models. Bisubstrate enzyme kinetic models include random Bi-Bi, and ordered Bi-Bi; and steady-state kinetic models such as Theorell-Chance mechanism, ordered Bi-Bi, ordered Uni-Bi, ordered Bi-Uni, ping-pong Bi-Bi, and random Bi-Uni. Examples of trisubstrate kinetic mechanism are ternary-complex mechanism, Ter-Bi mechanism, and Ter-Ter mechanism. Michaelis–Menten model (or Monod Kinetics model) is employed on the determination of initial rate of reactions. The initial rate of reactions is determined by plotting time profiles of decreasing substrate concentration or increasing product concentration, or increasing catalyst concentration. The MM kinetic mechanism assumes that binding of the substrate (S) to the catalytic sites of enzyme (E) forms the enzyme–substrate [ES] complex and generates product (P). At the same instance of formation of ES complex, the dissociation also occurs to free E. ES complex formation step is fast and reversible in nature while the production of P is slow and irreversible in nature. Type and nature of inhibition pattern can be deduced from graphical representation techniques, such as Lineweaver–Burk plot (or double reciprocal plot), Hanes Plot, and Dixon plot.

4.2 Monitoring of Sonication-Induced Structural and Morphological Changes in Enzyme

Previous studies on the ultrasound-assisted enzymatic synthesis have reported that on a broader picture ultrasonic waves can alter the structure of enzymes. These changes result in the enhancement of the kinetics of chemical processes. To give a brief understanding of how ultrasound waves help in enhancing the kinetics of the enzymatic process, one needs to monitor which changes are being made in the functional structure of enzyme. For investigating the impact of ultrasound, one can use various instrumental approaches that have been used in previous studies. These changes can be monitored using spectroscopic techniques (UV–Vis and fluorescence spectrometry), Fourier-transformed infrared spectroscopy (FT-IR), and Circular Dichroism spectroscopy. Using similar approach, Nadar et al. reported ultrasound treatment on *Candida antarctica Lipase B* (CALB) and lipase from *Thermomyces lanuginosus* (TL) using FT-IR analysis and intrinsic fluorescence analysis. The authors detected that ultrasonic irradiation increased the tryptophan moiety in both lipases. They also observed that the fraction of random coil in *CALB* was increased to 21.54% and in TL was 25.99%. Thus, authors concluded that the treatment of ultrasound induced the structural unfolding in both of the lipases, and the location of amino acid tryptophan and tyrosine side chains were also modified. While analyzing particle size distribution, authors observed that ultrasound treatment prohibited the agglomeration of lipase and increased the surface area thereby improving the catalytic activity. The enhancement of catalytic activity in TL and *CALB* lipase increased by 1.3-fold and 1.5-fold, respectively, due to cavitational effects of mild ultrasound frequency. The rate of degradation exhibits pseudo-first order behavior and the combination method was more effective than either sonolysis or enzyme treatments applied separately. The sonication phenomenon not only generates cavitation in liquids but also the formation of microjets, heating, and interfacial activity at interfaces of liquid–liquid, liquid–solid, and liquid–gas interfaces. In homogenous systems, it is observed that reaction occurs on two levels, one arises from the gaseous phase inside bubble termed as "heat point" and the second is associated with the thin layer of the liquid phase that surrounds this bubble. Whereas in heterogeneous systems, asymmetrical implosion of bubble causes the jet formation of liquid toward the solid surface that leads to localized destabilization. Also, destruction of bubble action build up micro-flow of liquids increasing the mass transfer that can enhance the chemical interaction between solid and liquid phase. For investigating the effect of sonication on lipase from *Burkholderia cepacia,* esterase assay was carried out with substrate as para-nitro phenyl palmitate. Next, lipase samples were screened in far-UV-CD spectra. Through the analysis, authors observed sonication did not affect the secondary lipase structure while in near UV region phenylalanine, tyrosine, and tryptophan it did made positional changes. Through recording the CD spectra, authors observed perturbation in tertiary structures that includes increment in band intensity in negative region. These effects occurred in microenvironments of aromatic residues, viz. tyrosine and tryptophan. Next, author conducted SEM analysis of ultrasonic-treated and non-treated lipase sample that showed morphological and structural alteration at the molecular level. In the ultrasonic-treated lipases, a fine layer on the surface of catalyst was seen which might have increased the catalyst surface area.

4.3 QM/MM Simulations to Understand the Molecular Mechanism of Enzymatic Reactions

The quantum mechanics/molecular mechanics (QM/MM) simulation is a hybrid technique in molecular simulation methods to study chemical process/reactions in reaction system involving biological macromolecules. The QM/MM integrates the accuracy of ab initio quantum mechanics calculations and speed of molecular mechanics methods. It was first introduced by Warshel and Levitt in the year 1976. These authors along with M. Karplus received Chemistry Nobel Prize in 2013 for "the development of multiscale models for complex systems". The hybrid QM/MM approach is being used to study the dynamics of biological macromolecules (DNA, RNA, proteins, or co-factor) in solution(s) (such as aqueous, organic, or mixed) since decades. In recent years, it has been found to be great tool to elucidate the molecular mechanism of enzymatic reactions. With the help of molecular docking simulation software (e.g., Autodock) researchers are able to map the binding site of substrates/inhibitors to the enzymes/proteins that assist in designing the desirable features in enzymes/proteins. Gaussian, a software for computational chemistry, allows scientists to predict energies, spectroscopic data (UV, NMR, IR, etc.), molecular structures, and advanced calculations by employing fundamental laws of quantum mechanics. It has been applied by researchers to predict very accurate chemical structures of a complex molecular which can further be utilized in other computational studies. Software for molecular dynamic simulation such GROMACS, NAMD, and AMBER assists mainly dynamics simulation of proteins, lipids, and nucleic acid in any liquid solvents. Figure 4 represents the molecular mechanism and steps involved in lipase-catalyzed esterification. This information is very much essential to understand the type of enzyme–substrate kinetics, which is ping-pong bi-bi in this case. The catalytic site consists of five active amino acids, viz. Ser, Gln, Thr, His, and Asp. In the general esterification reaction, the first substrate is acid which is being attacked by the nucleophilic site generated over Histidine and the first transition state is formed (TS-I in Fig. 4). Subsequently after structural rearrangements, the first product, water molecule, is released. This generates a tetrahedral intermediate (TI in Fig. 4) in the active site and modified enzyme consisting part of carboxylic acid (first substrate). This is followed by attack of second substrate and formation of second transition state (TS-II in Fig. 4). Second structural rearrangements are followed by release of second product (ester molecule), and native structure of active site is generated.

4.4 Amalgamation of Mathematical Modeling, Secondary Structural Analysis, and QM/MM Simulations

As represented in Fig. 5, the three amalgamated approaches can be represented as the three vertices of the solution triangle for unraveling the mechanistic aspects of ultrasound-assisted enhancement in reaction kinetics. These vertices are (1) estimation of secondary structural changes in enzymes, (2) computational techniques such as molecular docking, and (3) mathematical modeling to understand and explain kinetics of the reaction. It is known from the previous literature that ultrasound irradiation has impacts on the secondary structure composition of enzymes which can be estimated via techniques such as deconvolution of FT-IR spectra, CD spectroscopy, and Fluorescence spectroscopy. To explain the impacts of structural changes on reaction kinetics, one can perform molecular docking analysis that reveals the molecular mechanism of enzymatic reactions and product formation. Furthermore, mechanistic investigation by using mathematical models can elucidate the thermodynamic and macroscopic effects of ultrasound on reaction kinetics to make it faster.

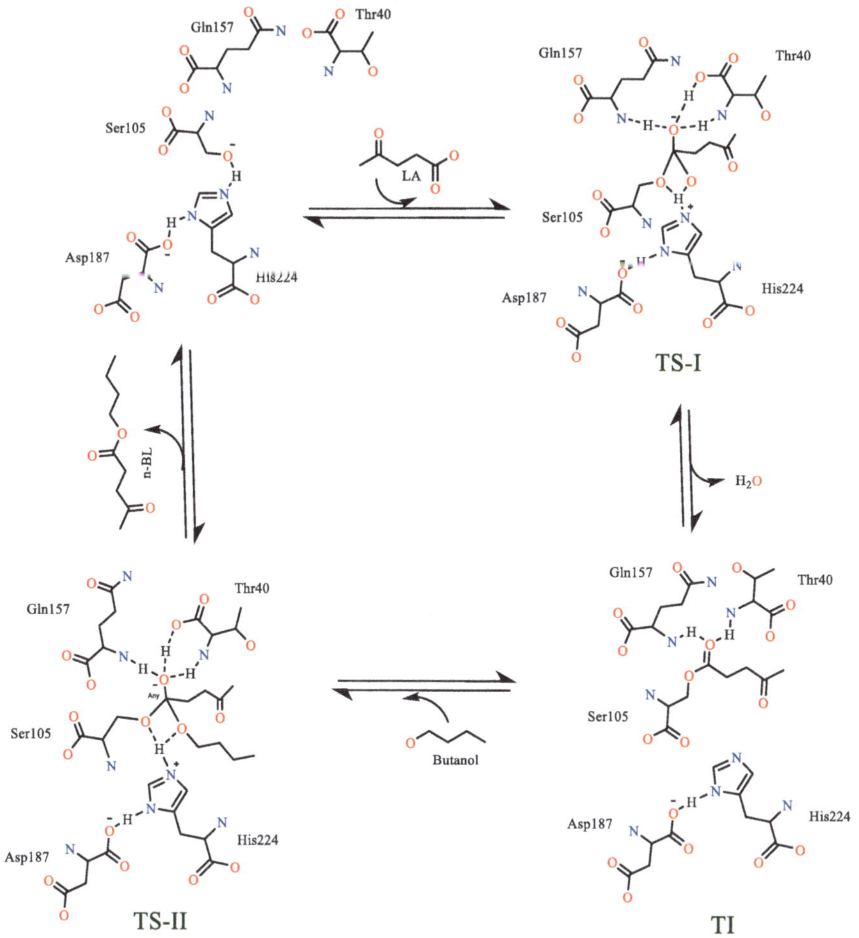

Fig. 4 Molecular mechanism and steps involved in lipase-catalyzed esterification deduced by molecular docking simulations (reproduced with permission from Kumar et al. [9])

5 Overview and Conclusions

In this article, we have proposed a methodology which is amalgamation of three simultaneous approaches (including both computational as well as experimental techniques), viz. quantum chemical simulation, mathematical modeling, and secondary structural analysis of enzymes to unravel the mechanistic aspect of enhanced reaction kinetics with the application of sonication. To explain the impacts of structural changes on reaction kinetics, molecular docking analysis can be helpful in revealing the molecular mechanism of enzymatic reactions and product formation. Furthermore, mechanistic investigation by using mathematical models can elucidate the thermodynamic and macroscopic effects of ultrasound on reaction kinetics to make

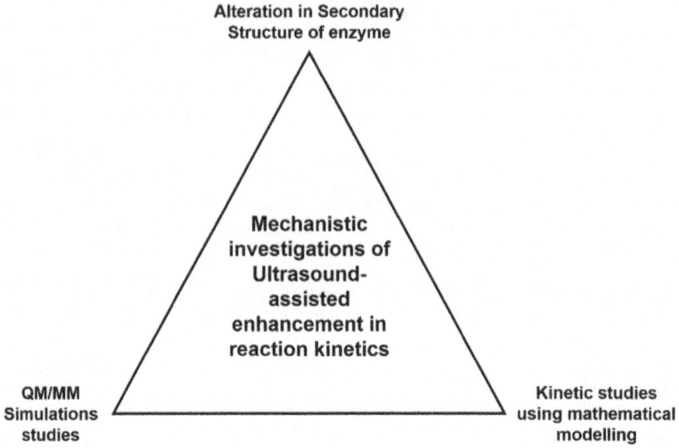

Fig. 5 Schematic representation for amalgamated approach for unraveling the mechanistic aspects of ultrasound-assisted enhancement in reaction kinetics

it faster. With this approach, researchers will be able to contribute to the knowledge in physiochemical effects of ultrasound waves in enzymatic reaction systems. This amalgamated approach may support future research for finding out the mechanism of sonication-induced product enhancement and similar processes.

Acknowledgements Mr. Karan Kumar is grateful to the Prime Minister's Research Fellowship provided by the Ministry of Education, Government of India.

References

1. P. Anastas, N. Eghbali, Chem. Soc. Rev. **39**, 301 (2010)
2. M.C. Bryan, P.J. Dunn, D. Entwistle, F. Gallou, S.G. Koenig, J.D. Hayler, M.R. Hickey, S. Hughes, M.E. Kopach, G. Moine, P. Richardson, F. Roschangar, A. Steven, F.J. Weiberth, Green Chem. **20**, 5082 (2018)
3. J. Li, J. Albrecht, A. Borovika, M.D. Eastgate, ACS Sustainable Chem. Eng. **6**, 1121 (2018)
4. R.A. Sheldon, ACS Sustainable Chem. Eng. **6**, 32 (2018)
5. S. Sen, J.E. Puskas, Molecules **20**, 9358 (2015)
6. N. Ran, L. Zhao, Z. Chen, J. Tao, Green Chem. **10**, 361 (2008)
7. C. Ortiz, M.L. Ferreira, O. Barbosa, J.C.S. dos Santos, R.C. Rodrigues, Á. Berenguer-Murcia, L.E. Briand, R. Fernandez-Lafuente, Catal. Sci. Technol. **9**, 2380 (2019)
8. Q.A. Al-Maqtari, W. AL-Ansi, A.A. Mahdi, **8**, 21 (2019)
9. K. Kumar, K. Roy, V.S. Moholkar, Process Biochem. **111**, 147 (2021)
10. R.S. Malani, S.B. Umriwad, K. Kumar, A. Goyal, V.S. Moholkar, Energy Convers. Manage. **188**, 142 (2019)
11. S.D. Shinde, G.D. Yadav, Appl. Biochem. Biotechnol. **175**, 2035 (2015)
12. M. Habulin, S. Šabeder, M.A. Sampedro, Ž Knez, Biochem. Eng. J. **42**, 6 (2008)
13. M.-M. Zheng, L. Wang, F.-H. Huang, P.-M. Guo, F. Wei, Q.-C. Deng, C. Zheng, C.-Y. Wan, J. Mol. Catal. B Enzym. **95**, 82 (2013)

14. R.S. Malani, V. Shinde, S. Ayachit, A. Goyal, V.S. Moholkar, Ultrason. Sonochem. **52**, 232 (2019)
15. R. Jia, Y. Hu, L. Liu, L. Jiang, H. Huang, Org. Biomol. Chem. **11**, 7192 (2013)
16. Q. Liu, G. Xun, Y. Feng, Biotechnol. Adv. **37**, 530 (2019)
17. R. Singh, M. Kumar, A. Mittal, P.K. Mehta, 3 Biotech **6**, 174 (2016)
18. L.C. Phukon, R. Chourasia, M. Kumari, T.K. Godan, D. Sahoo, B. Parameswaran, A.K. Rai, Biores. Technol. **309**, 123352 (2020)
19. D. Wang, L. Yan, X. Ma, W. Wang, M. Zou, J. Zhong, T. Ding, X. Ye, D. Liu, Int. J. Biol. Macromol. **119**, 453 (2018)
20. P. Joyce, C.P. Whitby, C.A. Prestidge, RSC Adv. **6**, 13802 (2016)
21. R. Bandikari, J. Qian, R. Baskaran, Z. Liu, G. Wu, Biores. Technol. **249**, 354 (2018)
22. R. Çakmak, G. Topal, E. Çınar, Appl. Biochem. Biotechnol. **191**, 1411 (2020)
23. D.G. Filho, A.G. Silva, C.Z. Guidini, Appl. Microbiol. Biotechnol. **103**, 7399 (2019)
24. N.R. Khan, S.D. Gawas, V.K. Rathod, Bioprocess Biosyst. Eng. **41**, 1621 (2018)
25. K.C. Badgujar, B.M. Bhanage, J. Mol. Catal. B Enzym. **122**, 255 (2015)
26. K. Kumar, H. Shah, and V.S. Moholkar, in *Optimization of Sustainable Enzymes Production* (Chapman and Hall/CRC, 2022)
27. K. Kumar, P. Patro, U. Raut, V. Yadav, L. Barbora, V.S. Moholkar, *Ultrasonic Enhancement of Lipase-Catalyzed Reactions: Mechanistic Investigation Using Molecular Docking Analysis* (Preprints, 2022)
28. Y.-C. Yin, H.-Q. Li, X.-S. Wu, Appl. Biochem. Biotechnol. **191**, 1670 (2020)
29. S. Cebrián-García, A.M. Balu, R. Luque, Front. Chem. **6**, 197 (2018)
30. L.E.S. Brink, J. Tramper, K.Ch.A.M. Luyben, K. Van't Riet, Enzym. Microb. Technol. **10**, 736 (1988)
31. S. Krimm, J. Bandekar, in *Advances in Protein Chemistry*, ed. by C.B. Anfinsen, J.T. Edsall, F.M. Richards (Academic Press, 1986), pp. 181–364
32. W.K. Surewicz, H.H. Mantsch, Biochimica et Biophysica Acta (BBA)—Protein Structure and Molecular Enzymology **952**, 115 (1988)
33. M. Balen, C. Silveira, J.M. Kratz, C.M.O. Simões, A. Valério, J.L. Ninow, L.G. Nandi, M. Di Luccio, D. de Oliveira, Biocatal. Agric. Biotechnol. **4**, 514 (2015)
34. Q. Carboué, M.-S. Tranier, I. Perraud-Gaime, S. Roussos, in *Microbial Enzyme Technology in Food Applications*, ed. by R.C. Ray, C.M. Rosell, 1st ed. (CRC Press, Boca Raton, FL : CRC Press, [2016] | Series: Food Biology Series | "A Science Publishers Book.," 2017), pp. 437–451
35. K. Doriya, N. Jose, M. Gowda, D.S. Kumar, in *Advances in Food and Nutrition Research* (Elsevier, 2016), pp. 115–135
36. V.S. Ferreira-Leitão, M.C. Cammarota, E.C. Gonçalves Aguieiras, L.R. Vasconcelos de Sá, R. Fernandez-Lafuente, D.M.G. Freire, Catalysts **7**, 9 (2017)
37. M. Martinelle, M. Holmquist, K. Hult, Biochimica et Biophysica Acta (BBA)—Lipids and Lipid Metabolism **1258**, 272 (1995)
38. B. Stauch, S.J. Fisher, M. Cianci, J. Lipid Res. **56**, 2348 (2015)

Analysis on Implication Viability of Three-Wheeler Electric Rickshaws Penetration in Indian Market

Harshendra N. Shet K and Vijayanand S. Moholkar

Abstract In today's world, global warming is a serious issue of concern due to its effects on earth's surface in various aspects. Under the Paris agreement, most of the countries agreed to set the emission reduction target and pursue their efforts to achieve this goal. Out of various causes for the global warming, CO_2 emissions by the transportation sector have contributed about quarter of the total global emissions. One promising solution to bring down the tailpipe emission from internal combustion engine vehicles is electric vehicles. Studies inferred that, though the tailpipe emission is absent in electric vehicles, it emits greenhouse gases indirectly through its source of primary energy, which is majorly coal-fired thermal power plants. Upon that the manufacturing of Li-Ion batteries used in electric rickshaws emits huge CO_2 emissions which depend on battery emission factor in the range of 121–250 $kgCO_2/kWh$. So, it is very much important to estimate and compare the CO_2 emission by electric vehicles against conventional vehicles before completely accepting the replacement of conventional vehicles by electric vehicles. In this chapter, we have thoroughly reviewed various literatures regarding the lifecycle analysis of electrical vehicles and to examine that a brief analysis on well to tailpipe CO_2 emission by three-wheeler rickshaw segment in Indian market is estimated. Our preliminary analysis *prima facie* shows that the penetration of three-wheeler electric rickshaws replacing the conventional internal combustion engine rickshaws does not bring down the CO_2 emissions to support the emission reduction target unless the primary source of energy is shifted to low greenhouse gases emitting renewable energy sources like solar or wind energy. Surprisingly, the results also showed that by increasing the thermal power plant efficiency to 40%, around 10.14% of CO_2 emission reduction is achievable, but this does not help to achieve the nation's emission reduction targets.

H. N. Shet K (✉) · V. S. Moholkar
School of Energy Science and Engineering, Indian Institute of Technology, Guwahati 781039, India
e-mail: harshendra_n@iitg.ac.in

H. N. Shet K
Department of Electrical and Electronics Engineering, Manipal Institute of Technology, Manipal Academy of Higher Education, Manipal 576201, India

V. S. Moholkar
Department of Chemical Engineering, Indian Institute of Technology, Guwahati 781039, India

V. S. Moholkar et al. (eds.), *Sustainable Energy Generation and Storage*,
https://doi.org/10.1007/978-981-99-2088-4_6

Keywords Greenhouse gases · Battery electric vehicles · Internal combustion engine vehicles · CO_2 emission · Payback period · Thermal power plant

1 Introduction

1.1 Background

Global warming is a serious matter of interest in today's world with its adverse effects on increase in the earths' surface temperature, upsurge in sea level, glacier melting, change in weather pattern, etc. Among the various sources of global warming, transportation sector plays a considerable role due to its high intense energy requirements and tailpipe emissions. International Energy Agency (IEA) says that about 8222 million tonnes of CO_2 emission take places due to transportation sector alone which is around 24% of the total global emissions. Whereas the Indian transportation sector emits lesser than global average, around 308 million tonnes of CO_2 is 13% of total emission in India [1]. This is due to the low vehicles ownership per capita in India compared to the global average. It is expected that the vehicle ownership in India to grow several folds in different segment by 2040 [2]. With the prediction of 50% rise in CO_2 emissions by 2040 in India, there is an utmost need to accomplish emission reduction targets to recover the quality of air and lessen global warming. One such solution to reduce CO_2 emission in transportation sector is electric vehicle (EV) which is gaining attention and acceptance as they remove tailpipe emissions in use phase.

Additionally, the Indian government is actively promoting the use of electric vehicles in the domestic market. It has introduced the National Electric Mobility Mission Plan (NEMMP) 2020, which will allow the Indian automotive industry to overtake other countries as the world's leading producer of electric vehicles and so aid in the nation's fuel security. India's earlier goal of having all vehicles electrified by 2030 has been modified to a more modest target of 20 to 30% EVs on the road by that time [3, 4].

1.2 CO_2 Emission by Electric Vehicles and Indian Power Scenario

By encouraging hybrid and electric vehicles across the nation, the Government of India's National Electric Mobility Mission Plan (NEMMP) 2020 intends to ensure national fuel security. From 2020 onwards, there is a sincere goal of selling 6–7 million electric and hybrid automobiles annually [4]. However, 57.3% of the electricity in India is generated by coal-fired thermal power plants. Due to India's heavy reliance on coal for energy production, the penetration of electric vehicles will result in higher electricity consumption, adding to the energy load on thermal coal power plants. There is no denying that when coal burns, hazardous byproducts are produced negating the reductions in tailpipe emissions [5]. Data from ministry of power website as given in the Table 1 describe the present Indian power scenario [6]. As per the latest data dated March 2021, the total installed capacity in India is 379.13 GW. Out of which 53% of the installed capacity is coal-powered thermal power plants. Sector-wise and fuel-wise installed capacity distribution is shown in Tables 1 and 2 as given.

As per the power generation data of 2019–20, out of 1598 TWh of total energy generated in India, 994 TWh was generated by coal-based thermal power plant which constitutes to 62.2% of total energy generation. Additionally, India produces coal of inferior quality compared to other nations, which causes the emission intensity to be nearly twice as high as the average for the world [5].

The electric vehicle energy demand has to be supplied by these thermal power plants, which emits approximately 1.6 kg eq. CO_2 to produce 1 kWh of electrical energy. Hence, this CO_2 emission indirectly contributes as emission by electric vehicle. Though the electric vehicles are found to be promising factors in reducing the global GHG emissions, in real they are shifting their use phase CO_2 emissions from vehicles tailpipe to power generation plants. The GHG emissions would be much worse for battery electric vehicles (BEV) than conventional internal combustion engine vehicles (ICEVs) if the generation of electricity is not decarbonized [7].

Table 1 Sector-wise installed capacity by March 2021

Sector	Installed capacity (GW)	% in total
Central	96,187	25.4
State	103,628	27.5
Private	179,315	47.3
Total	379,130	100

Source Ministry of Power 2021 [6]

Table 2 Fuel-wise installed capacity by March 2021

Fuel	Installed capacity (GW)	% in total
Total thermal	233.171	61.5
Coal fired	201.085	53
Lignite fired	6.620	1.7
Gas fired	24.957	6.6
Diesel fired	0.510	0.1
Hydro (Renewable)	46.209	12.2
Nuclear	6.780	1.8
RES*(MNRE)	91.154	24.5
Total	379.130	100

Source Ministry of Power 2021 [6]

When implementing, this circumstance is a source of worry for "switch to electric vehicle" in India. Therefore, the adequately decarbonized power generation mix must be enabled for deep reductions in GHG emissions from the transportation sector. In some geographical regions with high clean and renewable power penetration, it is found that BEVs offer a substantial decrease in the transport sector GHG emissions [8]. So, in this regard, to support reduction of GHG emissions due to EVs by the year 2030, India is ambitious to generate around 40% of its installed electric power capacity from non-fossil fuel sources [9].

1.3 Life Cycle Analysis

It is evident that estimation of CO_2 emission is important to adopt the penetration of electric vehicles over conventional IC engine vehicles. In recent years, life cycle analysis (LCA) studies have found more attention as they can provide more comprehensive view of CO_2 emissions of electric vehicles. Most of the recent studies were limited with estimation of CO_2 emissions of run phase or pump to wheel analysis only. To access the energy efficiency and energy life cycle of a BEV or FCEV, it is necessary to estimate the energy from raw materials to wheels to driving range [10]. The manufacture, fuel extraction, refinement, power generation and end-of-life stages of a vehicle should also be taken into consideration in addition to the actual operational phase [11].

India has undergone massive urbanization over the past 20 years, with major rise in population of communities that are economically categorized as middle-class and upper middle class. This has concurred with huge rise in the sales and use of motor vehicles, which has culminated in number of problems, such as traffic congestion, accidents, energy waste, noise and air pollution. Especially, air pollution is a daunting issue as it directly affects human health. As a solution to this issue, ICEVs are being phased out in favour of electric cars, and this transition is expected to enhance air quality by lowering vehicular pollution. In order to reduce greenhouse gas (GHG) emissions and energy usage, governments of developing and developed nations have pushed for the adoption of hybrid and electric cars.

This study focuses on estimation of primary energy consumption, life cycle costs of BEV, payback period, CO_2 emission from electric vehicle and IC engine. As it is known that per kilometre operating cost of electric vehicle is lesser compared to IC engine vehicles, the capital and maintenance cost of electric vehicle makes it necessary to estimate the cost involved in maintenance and replacement of battery, and how much time it takes to justify the decision of shifting to electric vehicle is economical. So, a payback period estimation can justify it. It is important to understand the viability of electric vehicle in Indian market which majorly depends on power generation by coal-fired thermal power plant. So, it is required to compare the CO_2 emission by electric vehicle and IC engine and to propose solutions to make feasible penetration of electric vehicles in the Indian market.

2 Review of Existing Literature

In this chapter, a broad literature survey on sustainability of electric vehicle in Indian market followed by a review on the specific topic of estimation of CO_2 emission by conventional ICEV in terms and BEV in terms of CO_2 emission by thermal power plant, techno-economic analysis on feasibility of BEV, etc. are discussed.

- Jain et al. discussed the feasibility of penetration of electric vehicles in Indian market in near future. Starting with different types/models of electric vehicles, they presented an analysis of energy consumption pattern as increasing number of conventional vehicles being replaced by electric vehicles in steps of 10% per year [3]. The trend of production of electric vehicles by 2030 has been predicted by using data on automobile production from the Society of Indian Automobile Manufacturer (SIAM). An analysis of the required battery specifications for the EVs manufactured by different companies has also been done. On the basis of average distance driven by each category vehicle, the annual energy consumption has been estimated. This study has also compared the cost saving patterns with conventional vehicles and EV. The overall conclusion of this study is that fixed cost of electric vehicles is likely to be higher (as contributed by high cost of battery); however, this may be offset by lower operating cost due to high cost of petroleum fuels.
- Nimesh et al. have evaluated the effect of switching the fleet from internal combustion engine vehicles (ICEVs) to electric vehicles (EV) on power generation sources, i.e. coal power plants, and its impact on the environment in India [5]. The implication viability of electric vehicle over internal combustion engine vehicle has been assessed with exergy analysis, considering the total emissions involved from well to wheel. This study showed that electric vehicles emit less CO_2 and CO, whereas SO_2 and oxides of nitrogen (NO_x) emissions (which occur from the thermal power plant) are higher. The degree of viable implication is mathematically derived which indicates the sustainable switch of ridership from IC engine vehicles to electric vehicles. The result indicates that the higher emissions of SO_x and NO_x with increasing use of electric vehicles can be reduced by using electricity produced through renewable sources for charging the batteries.

From 901.7 gCO_2/kWh in 2005 to 926 gCO_2/kWh in 2012, India's coal power stations' emissions increased. Compared to the 542 g and 533 g CO_2/kWh global averages for the same time period, this is significantly higher. The implementation of India's switch to electric vehicles raises questions about this scenario.

They have also discussed on the line diagram of petroleum well to end user which comprises of many stages. After extraction of crude oil, it is transported to refineries where it goes through a number of procedures, such as fractional distillation, to separate the petroleum fuel and extract the necessary products. Following that, the necessary product is delivered to the fuelling station. Figure 1 given shows the various steps involved in fuel supply from well to user. Every step consumes a certain amount of energy and generates emissions in the process.

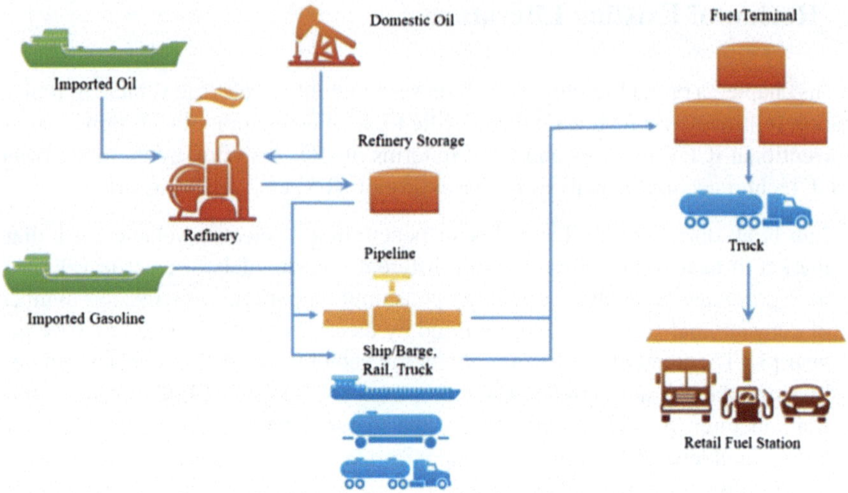

Fig. 1 Process flow diagram of fuel supply from well to user. *Source* [5]

- The study by Furuholt quantified the energy consumption and emissions involved in the production of gasoline and diesel [12]. Table 3 shows the well to tank GHG (equivalent CO_2) emissions involved in each process of production of petroleum-based fuel.

- Kennedy and Philbin [13] have presented techno-economic analysis of electric vehicles and compared them against hybrids vehicles and conventional vehicles. They have discussed on environmental considerations, technological challenges and potential for advancement, consumer influence on economic and environmental factors and life cycle cost comparisons. The study's findings indicated that

Table 3 Well to tank GHG emissions (equivalent of CO_2) of petroleum-based fuel production

Activity	CO_2 generated for gasoline (g/l)	CO_2 generated for diesel (g/l)
Crude oil extraction	41.064	58.286
Overseas transportation	32.016	35.512
Petroleum refining	298.932	102.676
Domestic transportation	14.268	14.282
Total	397.764	210.756

Source Furuholt [5, 12]

electric vehicles have a place in the future, but that further development is necessary before they can be categorically the best option in any given circumstance due to the complexity of electrical power supply, infrastructure requirements and full life cycle concerns.

- The main concerns with EVs and their effects on sustainability have been discussed by Dang et al. [14]. The Three-Pillar model including environmental impact, social impact and economic impact has been used for analysis, which gives an unbiased report of socio-economic impact of electric vehicles. They report that the Government is working with five approaches for the EV, which include (a) supporting early market, (b) shaping the necessary infrastructure, (c) securing the right regulatory and fiscal measures, (d) investing in the UK automotive capability and (e) preparing the energy sector. The study proposed three-pillar model, which gave an unbiased analysis on environmental, social and economic impact of electric vehicles. The study concluded that even with the widespread use of EVs in the future, there will undoubtedly be some unfavourable effects, such as battery disposal issues and what would truly happen to a smart grid if the harmonic pollution from EVs was not adequately addressed.
- Hara has presented a methodology in the form of a multidimensional techno-economic assessment diagram to broadly demonstrate the connection between assumed inputs and results [15]. This methodology was applied in order to analyse the life cycle costs and CO_2 emissions of hybrid vehicles (HVs) and electric vehicles (EVs). He also developed an eight-dimensional interactive diagram to show the relative advantages of HVs and EVs in the input space that consists parameters like HV fuel efficiency, EV energy efficiency, total mileage travelled, gasoline price, electricity price, battery price, gasoline CO_2 intensity and electricity CO_2 intensity. However, the study revealed complex correlations between presumptive inputs and results, even for simple functions, making it challenging to intuitively understand these linkages.
- Doluweera et al. have examined the energy and GHG emissions impact of EVs on Alberta's electricity and transport system. Utilizing the hybrid simulation model, they developed a new component to model EV fleets [16]. The impact of six scenarios with various assumptions regarding the rate of electric vehicle adoption and charging methods was examined using the modified model. The analysis exhibited that the adaptation of EVs could contribute to Alberta's 2030 emissions reduction target of 30% below 2005 levels. An examination of greenhouse gas emissions under the six scenarios showed that, comparing to the No EV scenario, by adapting EVs for passenger transportation, additional greenhouse gas emissions reduction is possible. For example, at 40% EV penetration, the total greenhouse gas emissions found to be only about 1 million tonne CO_2 eq, which is less than the No EV scenario emissions. A rise in EV penetration would result in a proportionate reduction in emissions from the passenger transportation sector, but increases the GHG emissions from the energy sector. The study concluded that it requires the implementation of electricity sector carbon management policies and programmes, which could lead to greater penetration of renewable sources of electricity generation to lower the GHG emission.

- Pareek et al. have discussed the overloading on the grid, charging time of batteries and traffic congestion due to more waiting time at electric vehicle charging stations (EVCS) [17]. To get beyond these challenges various charging technologies like smart grid technology, renewable energy technology, vehicle-to-grid technology and intelligent transport system were discussed. Charging techniques like constant current charging schemes, constant current constant voltage charging schemes and multistage charging schemes were also covered in order to improve the dependability and efficiency of the entire electrical transportation system. Out of these, multistage charging is discovered to be the most popular method of fast charging without deteriorating the battery. Authors also discussed as how to set up a charging station at home utilizing the directives provided by India's Ministry of Power and Ministry of Housing.
- Falcão et al. have presented a comparative study between an EV and medium-duty diesel ICEV using a standard drive cycle in urban driving conditions [18]. They have evaluated the performance constraints like electrical energy consumption and CO_2 emissions from power generation for EV. And for the conventional vehicle, tailpipe CO_2 emissions and energy consumed from fuel consumption and heating value. To undertake an economic viability assessment in terms of payback duration and net present value (NPV), they created five scenarios. The acquired results demonstrated that the electric vehicle's CO_2 emissions were 4.6 times lower than those of the diesel vehicle. However, the economic research showed that the EV's viability is threatened, mostly as a result of the imported parts' unfavourably high exchange rates. The calculated payback period of the EV is 13 years of operation in the best-case scenario and without taking revenue from commercial use.
- Neubauer et al. described a techno-economic analysis of second use of batteries taking into consideration the effects of battery degradation in both automotive and grid service, repurposing costs, balance-off system costs, the value of aggregated energy storage to commercial and industrial end users and competitive technology [19]. They said that the expensive price of batteries currently prevents plug-in electric cars (PEVs) from rapidly penetrating the market. The high cost of batteries also prevents the use of grid-connected energy storage, which might improve the grid's dependability, effectiveness and cleanliness. They suggested employing the battery in other secondary applications where it would still perform well enough to meet the demands of other energy storage applications. This will help to reclaim a portion of the battery cost after the battery is retired from electric vehicle. The lifetime value of the battery is extended by deriving additional services and income from it in a post-EV application, lowering the overall cost of energy storage systems for both primary (automotive) and secondary (grid) consumers.
- Kobashi and Yarime have mentioned that the current development in electric vehicles gives us a chance to boost residential solar photovoltaic (PV) self-consumption at a significantly lower additional cost [20]. With an eye towards 2030, they conducted an economic analysis of residential PV systems coupled with electric vehicles (V2H: Vehicle to Home) at Japanese households, incorporating the cost projections of these technologies in the future. It is observed that

a system that consists of PV and an EV is already cost-competitive with the use of electricity from grid and a gasoline vehicle in 2018. By 2030, combination of PV + EV would significantly improve the economics of energy for Japanese families, lowering yearly energy prices (electricity and gasoline) by as much as 68 percent and reducing the carbon footprint of the domestic energy system by 92 percent. Additionally, it is discovered that the PV + EV system is significantly more cost-effective than a PV only or PV + battery system because the large battery of the EV may be used with little additional expense.

- In Kyoto, Japan, and Shenzhen, China, Kobashi et al. have conducted an environmental techno-economic study of residential PV systems with batteries or EVs allowing charging and discharging (Vehicle to Home: V2H), considering the projected costs of these technologies by 2030 [21]. It is found that "PV + EV" (V2H) becomes highly cost-effective towards 2030 in comparison with "PV only", "PV + battery" and "EV charge". Results obtained concluded that "PV + EV" has the highest CO_2 emission reduction potential across all the technology combinations considered, owing to EV charging of carbon-free electricity from PV. They have suggested that, to facilitate a rapid household decarbonization, policymakers should reinforce policies to enhance the penetration of combined technologies of "PV + EV" (V2H) towards 2030 EVs can play a crucial role as an energy storage for the penetration of variable renewable energy (VRE), in addition to assisting in the reduction of CO_2 from the transportation sector.
- Hayajneh et al. proposed a new design of energy storage systems with battery accompanying wind mills from which the stored energy can be used for both stationary (backup) and mobile (electric vehicle) applications [22]. As their previous work focused on stationary applications, this work envisions a future in which transportation system is electrically powered. They proposed that electric vehicles get refuelled at charging stations that operate on batteries delivered by electric trucks from the wind farm energy storage systems. Based on the real-world situation of the Chapman Ranch wind farm and related data from ERCOT, a battery dispatch model is created and applied to scenarios with and without electric vehicle charging stations. According to the techno-economic study, the energy storage system's profitability index or investment appeal is increased by fusing stationary and mobile applications.
- Schetinger et al. have simulated the economic viability of electric vehicles (EVs) and renewable energy charging integration at a university campus by using Homer energy simulation software to determine the optimal solution for the power loading, energy dispatch and economic feasibility of each electricity source [23]. They have considered three scenarios: case 1 is the baseline that provided the electric parameters considering a grid without renewable energy integration; case 2 is the addition of solar and wind systems supplying power to the grid and EVs, and case 3 is the same conditions as case 2, but EVs operate on vehicle-to-grid (V2G) scheme supplying energy to grid. The obtained economic output possibilities were categorized based on the lowest net present value (NPV). The V2G mechanism taken into consideration in case 3 provided circumstances that reduced the system's operating costs, calculating the lowest NPV among the cases

examined. Case 2's simulations showed higher costs as a result of the system's functioning with the additional local load needed to power the EVs. In both case 2 and case 3 scenarios, EVs aided the renewable energy sources' penetration on the system; however, the operation of EVs on a V2G mechanism contributed to the highest rate of renewable energy penetration.

- Laurischkat and Jandt have created a techno-economic model of energy and mobility solutions based on system dynamics that illustrates the technical synergies between the three and the resulting costs for mobility and energy [24]. The model user can alter the customer requirements, environmental considerations and technical design possibilities through parameter modifications to obtain an instantaneous response regarding the change in costs. The study concluded that the simulation studies result with three different results. First, customers who rarely travel on highways and have a photovoltaic system will initially find electric automobiles to be economically appealing. Second, technological parameters of solar systems and battery storage that are cost-effective are significantly impacted by electric mobility. Third, the economic implications resulting from the technological collaborations can be used for future marketing programmes of municipal utilities. Additionally, the system dynamics model serves as the foundation for further investigation into the problem of varying electric prices, charging an electric vehicle at work using a photovoltaic system and connecting smart metres, electric vehicles, photovoltaic systems and battery storage systems via the internet to exchange information and record data.
- Qiao et al. have evaluated the CO_2 emission from manufacture phase of both EV and ICEV and compares the results [25]. Their obtained results revealed that CO_2 emission from EV is 14.6 tonne which is 59% higher than the level of ICEVs 9.2 tonne. Li-Ion batteries, traction motor and electronic controllers in EV are the major reasons behind this. CO_2 emission can be reduced by improving battery manufacturing technology and using recycled steel and aluminium.
- Kawamoto et al. have reported that due to additional emission by battery production for BEV, CO_2 emissions by BEV are much higher than an ICEV [11]. However, considering the regions with higher renewable energy mix, operational CO_2 emissions over a longer lifetime driving distance reduces for BEV compared to ICEV. But, lifetime driving emission estimation has to consider the replacement of battery, which shows the emissions of ICEV is smaller compared to that of BEV. In this study, the entire life cycle emission estimation was calculated in five different phases: vehicle production, fuel/electricity production, vehicle usage, maintenance and end-of-life as shown in Fig. 2. The study excludes the disposal and recycling of waste materials.

The total lifetime CO_2 emission estimated for BEVs and ICEVs in different regions shows that the CO_2 emission is dependent on the driving distance. The emissions by BEVs are lesser compared to ICEVs when the distances are longer. But, with the replacement of battery at 160,000 kms in BEV would alter the emissions level resulting ICEVs being better than them. Life cycle CO_2 emissions estimated for different countries is as shown in Fig. 3.

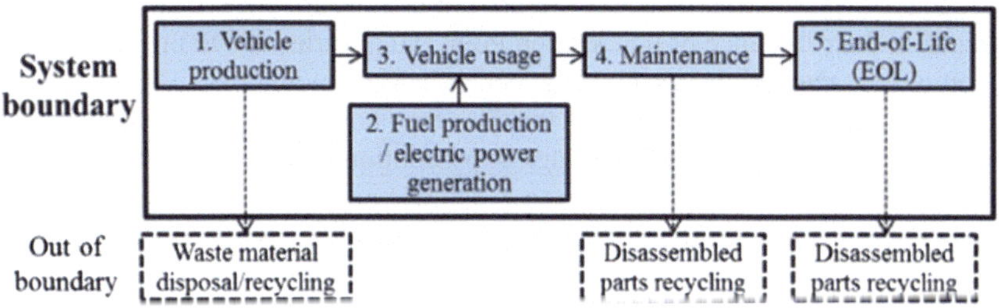

Fig. 2 Life cycle of electric vehicle. *Source* [11]

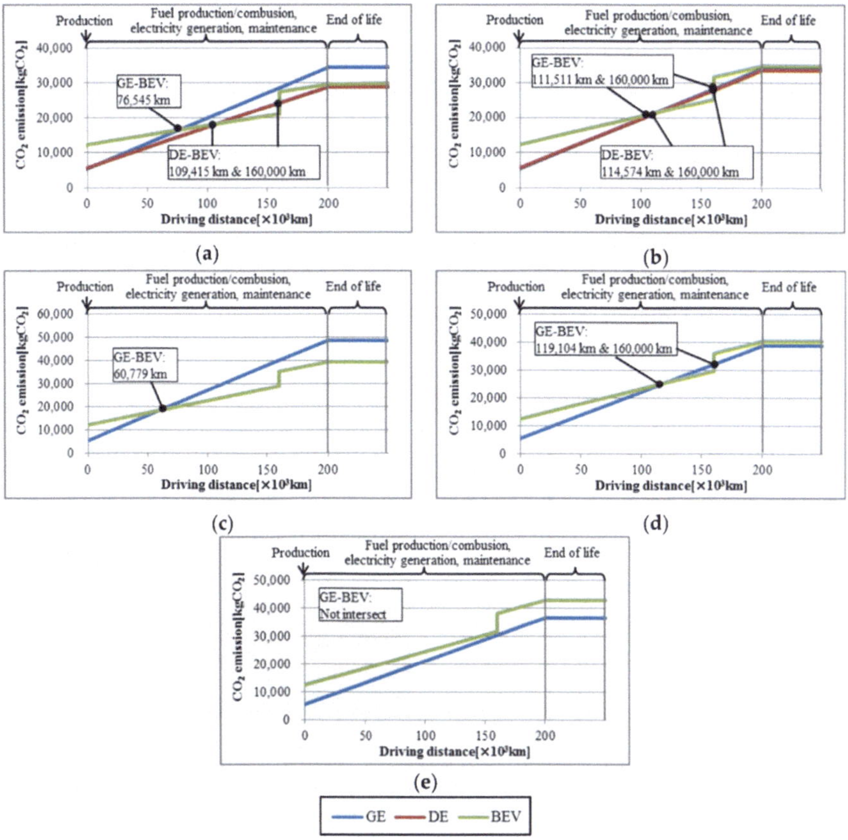

Fig. 3 Life cycle CO_2 emissions estimated for different countries. *Source* [11]

Though the production phase CO_2 emission is more for BEV than ICEV for all the regions, their Distance of Intersection Point (DIP) varies due to the difference in CO_2 emission factor of electric power generation. The amount of CO_2 emissions per unit battery capacity is represented by the CO_2 emission factor, which was estimated using (1) raw material extraction through to production of a battery system; (2) each detailed process of battery production (e.g. cathode production, cell assembly, pack assembly); (3) the lithium-ion battery included either mainstream cathode described as lithium nickel–manganese–cobalt oxide (NMC) cathode or lithium iron phosphate (LFP) cathode types. It was also observed that the battery emission factor (121–250 $kgCO_2/kWh$) plays an important role in emission value. Range of battery emission factor is described in Fig. 4. The average value considered is 177 $kgCO_2/kWh$.

- Rohr et al. have established a dynamic, parameter-based end-of-life (EoL) model which will help to analyse the economy of battery end-of-life value chains and estimate the residual value of the battery [26]. Their study showed that EV batteries are characterized by a limited lifetime based on the steadily decreasing ability to fulfil the minimum capacity and power requirements for the automotive application. Lack of knowledge on end-of-life (EoL) value chains, related interdependencies and dynamic circumstances make it difficult to determine the financial outcome and followed way of the EoL-options. It is concluded that the cost–benefit and net present value-based model provided is effective at filling the research gap. For the assumed baseline scenario, the survey showed that all EoL strategies recycling, remanufacturing and second-life possess economic potential in Germany.

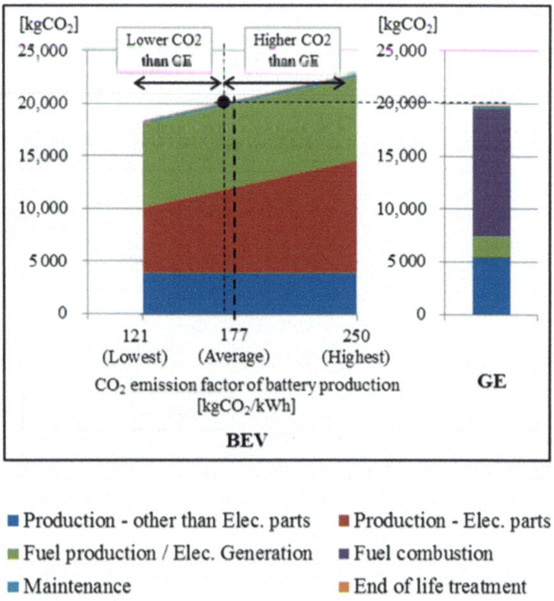

Fig. 4 Effect of battery emission factor on CO_2 emission. *Source* [11]

The method makes it possible to monitor the deterioration of the batteries in electric vehicles and forecast its future evolution for vehicle use as well as potential second-life scenarios. As a consequence, ageing uncertainties can be lessened.

- Abas et al. have examined the viability of introducing electric vehicles into the Brunei's market using life cycle cost analysis and also identified the dominant factors that influence its feasibility [27]. Though the data used was from local scenarios, the methodologies adopted can be applied for analysis of other markets. Their study showed that EV is presently still expensive when compared to conventional ICEV and hybrid electric vehicles (HEV) as its procurement cost contributes much to its life cycle cost (LCC). They have proposed that government should introduce subsidy of USD$4100 and increase the gasoline price to USD$0.70/litre to promote EVs in the market and make other vehicles less desirable. The study also suggested that in order to make EVs more environmentally competitive with conventional vehicles, it is necessary to consider cleaner renewable energy sources and enhance power plant efficiency.
- Morse discussed on the challenges associated with recycling old batteries in the near future when manufacturers anticipate producing millions of electric vehicle (EV) batteries over the following few decades [28]. It is challenging to develop a successful recycling system since batteries vary greatly in their chemical composition and structure. As a result, battery manufacturers prefer to use freshly mined metals over using recycled resources. According to their survey, recyclers typically aim to recover valuable cathode metals like cobalt and nickel. But because of the small quantities, the metals are like needles in a haystack: hard to find and recover. To extract those needles, recyclers rely on two methods, known as pyrometallurgy and hydrometallurgy. Both procedures, according to studies, generate a lot of trash and release greenhouse gases. Additionally, the business model might be unstable because battery manufacturers are attempting to move away from recovering cobalt, which is a rather costly metal and on which most companies rely to remain afloat. In such case, recyclers could be left trying to sell dirt hills. The study says researchers are advising EV and battery manufacturers, to start designing their products with recycling in mind. After China started holding EV makers accountable for ensuring batteries are recycled in 2018, The Blade Battery came into being. The nation primarily uses pyro- and hydrometallurgical techniques to recycle more lithium-ion batteries than the rest of the globe combined.

3 Methodology

In this section, the life cycle cost estimation of electric vehicles is estimated by analysing the capital and operating cost of electric vehicle, and payback period is determined. Also, CO_2 emission is estimated for both electric and IC engine vehicles and analysed the feasibility of penetration of electric vehicle in Indian market. Lighter segment electric vehicles consist of four-wheeler passenger, four-wheeler commercial, three-wheeler rickshaws and two-wheelers. In India, the majority of mobility demands are satisfied by informal transportation services, also known as Intermediate Public Transport (IPT) or paratransit systems, as autorickshaws, taxis, minibuses, etc. In most developing cities, there is insufficient access to public transportation; therefore, intermediate public transportation fills the gap. The autorickshaw is the most often used form of intermediate public transportation. It is a common option in South Asian nations including India, Bangladesh, Indonesia, Pakistan and Thailand since it is an economical, practical and an essential form of transportation [29]. So, in our analysis three-wheeler rickshaw segments is considered. The estimation of capital and operating costs of conventional rickshaw and electric rickshaw is made, and payback periods are determined. Also, estimation of CO_2 emission per kilometre is made for both rickshaws, and the feasibility of penetration of electric rickshaw in the Indian market is been analysed.

3.1 Life Cost Estimation

3.1.1 Wheeler Rickshaw Sales in India

Autorickshaws play a crucial part in Indian cities, albeit their use varies from one city to the next. Despite having a low vehicle share, three-wheelers remain the preferred means of transportation in Indian cities. Autorickshaws serve as the primary mode of public transportation in cities without any government-provided services, such as bus or rail-based transportation. Autorickshaws function as both a feeder to and an intermediary type of public transportation in other cities with established public transportation services. As per a report from TERI, three-wheeler rickshaws contribute about 3% of total CO_2 emissions emitted in India [30]. In order to estimate the impact of shifting from conventional IC engine rickshaws to EV rickshaws, it is the first step to identify the total number of three-wheeler rickshaws on the road in last 15 years. We have considered 15 years data as it is the nominal lifetime of a rickshaw [31]. The details of three-wheeler domestic sales are captured from the report shared by Society of Indian Automobile Manufacturers (SIAM). The year-wise sales trend is plotted as shown in the Fig. 5. The data shows that approximately 7,500,000 units of autorickshaws are sold in India in last 15 years from 2005 till 2020 [32].

3.1.2 Specifications of the Rickshaws

In the analysis, Bajaj Compact RE (BS6) rickshaw is selected under conventional IC engine rickshaw, and Mahindra TREO is selected under electric rickshaw. Few important data required for the analysis is fetched from the technical specification sheet of both rickshaws and listed in Table 4.

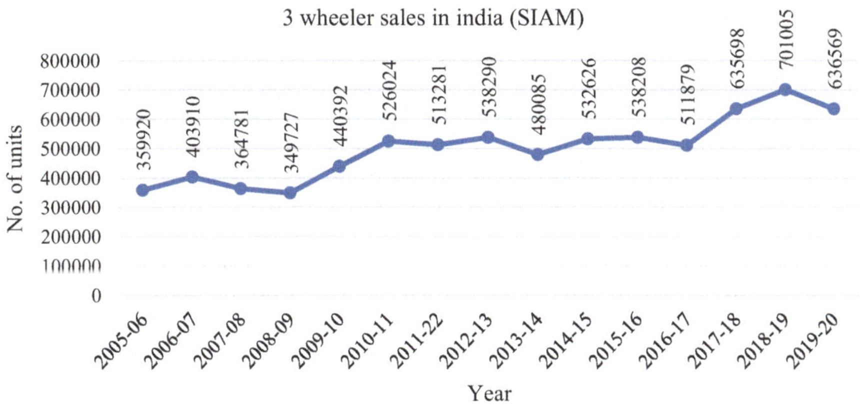

Fig. 5 Three-wheeler rickshaws domestic sale data from 2005 to 2020 from SIAM annual report. *Data Source* [32]

Table 4 Data sheet specifications of both rickshaws

Bajaj Compact RE (BS6)		Mahindra TREO (EV)	
Seating capacity	3	Seating capacity	3
Kerb weight	362 kg	Kerb weight	377 kg
Top speed	70 kmph	Top speed	55 kmph
Fuel efficiency	35 kmpl	Range	130 km
Battery	12 V, 32 Ah	Battery	48 V
Engine displacement	236 cc	Battery capacity	7.37 kWh
Peak power	7.6 kW	Charging time	4 h
Peak torque	19.2 Nm	Drive train peak power	8 kW
Tank capacity	8 L	Drive train peak torque	42 Nm
Capital cost	2.5 lakhs	Capital cost	3.5 lakhs

Source [33, 34]

3.1.3 Estimation of Operating Cost

From the data sheet in Table 4, it is found that capital cost of electric vehicle is high as the Li-Ion battery and motor which turns out to be expensive compared to IC engine rickshaw. On-road capital cost of Bajaj Compact RE is 2.6 lakhs, whereas Mahindra TREO costs 3.5 lakhs without any subsidy. Few assumptions like average distance travelled per day, number of annual working days, cost of petrol, cost of per unit electricity, etc. are being assumed to estimate the operating cost and determine the payback period. Assumptions made are as listed below.

- Distance travelled/day 70 km
- Cost of electricity 6 Rs/kWh
- Cost of petrol 93 Rs/litre (as per May 2021 data)
- Efficiency of battery charger 90%
- No. of days of working annually 300 days/year

Considering fuel efficiency of Bajaj Compact RE to be 35 kmpl, in a day to travel an average distance of 70 kms, 2 L of petrol is required. Considering an approximate cost of per litre of petrol to be Rs. 93, total cost of fuel per day to travel 70 kms is Rs. 186/-.

Therefore, operating cost of Bajaj Compact RE per kilometre is found to be Rs. 2.66/-.

Similarly, Mahindra TREO uses a battery of 7.37 kWh rating which operates at 48 V. So, the Ah rating of the battery is found to be 153 Ah. Considering the efficiency of battery charging adapter to be 90%, energy required to fully charge the battery is 8.19 kWh. This fully charged battery can last for about 130 kms. So, energy consumed to travel one kilometre distance is 0.063 kWh.

Considering the per unit (kWh) cost of electricity is Rs. 6, total cost of electricity bill per day to travel 70 kms is Rs. 26.46/-.

Therefore, operating cost of Mahindra TREO electric per kilometre is found to be Rs. 0.38/-.

It is found that the operating cost per kilometre of Mahindra TREO electric rickshaw is much lesser than Bajaj Compact RE. Approximately 85% of operating cost saving can be made. But the capital cost of electric vehicle is higher than conventional rickshaw. So, a payback period analysis is required to know how the excessive capital investment can be claimed by running electric rickshaw. To estimate the payback period, total mileage for 300 operating days per year is considered with 70 kms average mileage per day and total annual distance travelled it is found to be 21,000 kms. In order to make the analysis more accurate, evaluation of annual maintenance cost and depreciation cost are found to be important.

3.1.4 Estimation of Annual Maintenance Cost

The battery of Mahindra TREO has a warranty for 3 years and life span of 5 years; hence, investment is required on replacement of battery after 5 years. The rating of battery used in Mahindra TREO is of 48 V, 153.5 Ah. In market, approximate cost of this battery is around Rs. 120,000/-. As the electric rickshaws are maintenance free with less mechanical parts replacements compared to IC engine rickshaws, additional maintenance cost of electric rickshaws is assumed to be 10 paise/kilometre.

Total maintenance cost for 5 years is found to be $= (5 \times 21{,}000 \times 0.1) + 1{,}20{,}000$ $=$ Rs. 130,500/-So annual maintenance cost of Mahindra TREO will be Rs. 26,100/-.

For Bajaj Compact RE, as there will be replacement of mechanical parts after few thousands of kilometres over 5 years of period, approximate annual maintenance cost is assumed to be 75paise/kilometre. So, the Annual maintenance cost of Bajaj RE will be Rs. 15,750/-.

3.1.5 Depreciation Cost Estimation

Both rickshaws' depreciation costs are estimated using the straight-line depreciation approach. Until it reaches its salvage value, the value of an asset is consistently decreased during each month. The most popular and simple depreciation technique for dividing up the cost of a capital item is straight-line depreciation. The formula for calculating it is as simple as dividing the asset's cost by its useful life after deducting any salvage value.

Depreciation Cost/year = (Capital Value − Salvage Value)/Life Span

As per the revised report of Hakeem committee in 2017, in the Sect. 7.4.2, it is considered 15 years as the useful life of the auto rickshaw [31]. After 15 years of usage the auto rickshaw will be left with essentially its scrap value which can be assumed at 5% of the capital value.

So, the salvage value of Bajaj Compact RE is Rs. 13000/- and for Mahindra TREO is Rs. 17,500/-.

Depreciation cost/year of Bajaj Compact RE = (260,000 − 13,000)/15 = Rs. 16,466/-.

Depreciation cost/year of Mahindra TREO = (350,000 − 17,500)/15 = Rs. 22,166/-.

3.2 *Estimation of CO_2 Emission*

In order to estimate the CO_2 emission by both rickshaws, certain standard data are to be considered or assumed and are listed below.

- 1 L of petrol emits 2.35 kg CO_2
- 1 L of diesel emits 2.69 kg CO_2
- 1 kwh of electricity generation in thermal power plant 1.15 kg CO_2
- Efficiency of thermal power plant as 25%
- Efficiency of electricity transmission as 90%
- Calorific value of burning 1 kg of coal 32.77 MJ/kg
- CO_2 emission for 1 L of petrol production to user process is 398 g/litre.

To travel a distance of 70 kms per day Bajaj Compact RE consumes about 2 L of petrol. So annually for 300 days around 600 L of petrol is consumed. If 2.35 kg of CO_2 is emitted by burning 1 L of petrol, then annually Bajaj Compact RE emits about 1.41 tonnes of CO_2 from tank to tailpipe.

But, the crude oil is extracted from the ground and refined to create desired products, such as gasoline, diesel, kerosene, naphtha, LPG and heavy fuel oil, among others. Following crude oil extraction, it is transported to refineries where it goes through a number of procedures, such as fractional distillation, to separate the petroleum fuel and extract the necessary components. Each process step uses a certain amount of energy and emits CO_2. So, 398 g/litre of CO_2 emission has to be considered while estimating the well to wheel emissions. In this case, for 600 L of petrol consumption 238.80 kg of CO_2 emission is to be added. So, total CO_2 emission from well to tailpipe by Bajaj Compact RE annually is 1.648 tonnes of CO_2.

The electric rickshaw Mahindra TREO though it doesn't emit any emission from tailpipe but the energy consumed by it has to be supplied from the power grid which is majorly supplied by coal-fired thermal power plant. Thermal power plants emit CO_2 by burning coal. Hence, this CO_2 emission indirectly contributes as emission by electric vehicle.

The battery used in Mahindra TREO is of 7.37 kWh rating. The battery charger used is assumed to have 90% efficiency. So, 8.19 kWh of electric energy is required to once charge the battery to full.

For daily mileage of 70 kms, the annual energy consumed for 300 days will be 1.322 MWh which is to be supplied by thermal power plant. Considering electricity transmission efficiency to be 90% the actual energy to be supplied by thermal power plant will be 1.470 MWh.

Assuming nominal thermal power plant efficiency of 25%, total input energy required by burning coal is 5.879 MWh which is 21165 MJ. Considering calorific value of burning 1 kg of coal is 32770 kJ/kg, mass of coal required to produce 5.879 MWh electricity is 646 kg. Atomic weight of carbon is 12 g/mol and carbon dioxide is 44 g/mol. Then 1 kg of carbon produces 3.67 kg of carbon dioxide.

So, the total amount of carbon dioxide emitted by burning 646 kg of coal is 2.370 tonnes which is indirectly emitted by Mahindra TREO annually by consuming 1.470 MWh of electricity.

Comparing CO_2 emissions by both rickshaws, surprisingly it is found that Mahindra TREO emits 721.5 kg (i.e. 43.76%) more than conventional Bajaj Compact RE.

4 Results

4.1 Payback Period

Table 5 summarizes the different cost components associated with both rickshaws. Total annual mileage considered is 21,000 kms. Total annual running cost is estimated which includes annual fuel cost, maintenance cost and depreciation cost.

Table 5 Payback period estimation

	Bajaj Compact RE	Mahindra TREO
Annual mileage	21,000 kms	21,000 kms
Capital cost	Rs. 260,000/-	Rs. 350,000/-
Running cost per km	Rs. 2.66/-	Rs. 0.38/-
Annual fuel cost	Rs. 55,860/-	Rs. 7980/-
Annual maintenance cost	Rs. 15,750/-	Rs. 26,100/-
Annual depreciation cost	Rs. 16,466/-	Rs. 22,166/-
Total annual running cost	Rs. 88,076/-	Rs. 56,246/-
Net expense after 1st year	Rs. 348,076/-	Rs. 406,246/-
Net expense after 2nd year	Rs. 436,152/-	Rs. 462,492/-
Net expense after 3rd year	Rs. 524,228/-	Rs. 518,738/-
Payback period	2.84 years	

Total running cost/year of Bajaj Compact RE is Rs. 88,076/- and of Mahindra TREO is Rs. 56,246/- which shows that there is net annual saving of Rs. 31,380/- . Net expenses at the end of first, second and third year calculated show that net expenses of Mahindra TREO were higher at the end of first two years. But after third year, net expenses Mahindra TREO are less than that of Bajaj Compact RE. With this using interpolation, we can estimate that **payback period is 2.84 years**.

MESCOM Electricity Tariff for electric vehicle charging station under LT6c category charges per unit of electricity at Rs. 6/- [35]. Considering this price, the entire cost analysis is made. But it is of obvious understanding that in future with more electric vehicle penetration the electricity charge per unit will be tariffed under commercial sector which will be at least 2 to 3 times of domestic price. In that case, the payback period estimated will get increased by multiple folds.

4.2 CO_2 Emissions with Penetration Electric Rickshaws

With the results obtained in Sect. 3.2, this analysis is extended to understand the CO_2 emission with percentage penetration of electric rickshaw into Indian market by 2030 in steps of 10%. From SIAM Automobile Domestic Sales Trends (2021), considering the yearly increment in sales up to 2030 it is expected that approximately 7,500,000 rickshaws will be in Indian roads [36]. With percentage penetration of electric rickshaws in steps of 10% per year, CO_2 emission per year is estimated and results are as shown in Table 6 and Fig. 6.

Table 6 Tank to tailpipe CO_2 emission in Mt/year for penetration of electric rickshaws in steps of 10% using coal energy-based power generation

Electric rickshaw penetration (%)	No. of electric rickshaws	CO_2 emission/year (million tonne/year) (EV)	No. of ICEV rickshaws	CO_2 emission/year (million tonne/year) (ICEV)	(EV + ICEV) Total CO_2 emission (million tonne/year)	Increase in CO_2 emission (million tonne/year)
0	0	0	7,500,000	10.58	10.58	0
10	750,000	1.60	6,750,000	9.52	11.12	0.54
20	1,500,000	3.20	6,000,000	8.46	11.66	1.08
30	2,250,000	4.80	5,250,000	7.40	12.20	1.63
40	3,000,000	6.40	4,500,000	6.35	12.74	2.17
50	3,750,000	8.00	3,750,000	5.29	13.29	2.71
60	4,500,000	9.60	3,000,000	4.23	13.83	3.25
70	5,250,000	11.20	2,250,000	3.17	14.37	3.80
80	6,000,000	12.80	1,500,000	2.12	14.91	4.34
90	6,750,000	14.40	750,000	1.06	15.46	4.88
100	7,500,000	16.00	0	0.00	16.00	5.42

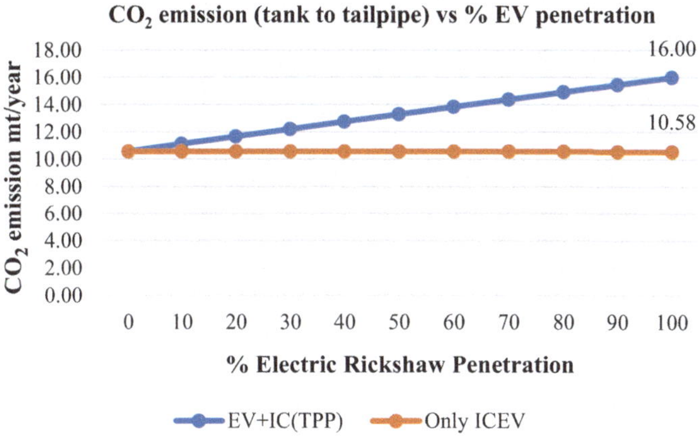

Fig. 6 Tank to tailpipe CO_2 emission versus %EV rickshaws penetration

Firstly, CO_2 emissions from tank to tailpipe is estimated and tabulated as shown in Table 6. It is observed that as the penetration of electric rickshaws increases, CO_2 emission also increases. With 100% electric rickshaws penetration CO_2 emissions increases by about 5.42 million tonne per year (51.30%). This analysis can be further made more realistic by estimating well to tailpipe emissions considering the emissions due to fuel production, refining and transportation to user phase for IC engine rickshaws and considering the electricity transmission efficiency for electric rickshaws. The obtained results are tabulated as shown in Table 7.

The above obtained results in Fig. 7 for 100% penetration of electric rickshaws still show a huge increase in the CO_2 emissions by, i.e. 5.41 million tonne/year. As the emissions by IC engine rickshaws increased due to its emissions during fuel production and processing, the increase is 43.76% which lower compared to previous case.

However, this is due to the poor efficiency of thermal power plant which was considered as 25%. Given the numerous applications for electricity worldwide including electric vehicles, increasing the efficiency of thermal power plants would be a useful strategy for lowering overall energy consumption and safeguarding the environment [37]. If the thermal power plant efficiency increases to 35–40% considerable amount of reduction in CO_2 can be expected. By increasing the efficiency of thermal power plant, CO_2 emissions are estimated, tabulated and plotted in Table 8 and Fig. 8 respectively as given.

The above results show that with increase in thermal power plant efficiencies, well to tailpipe emissions by electric rickshaws can be lesser than IC engine rickshaws when plant efficiency is more than 35%. At 40% plant efficiency, the CO_2 emissions from 100% electric rickshaw penetration is 11.11 million tonne CO_2/year which is 10.14% of reduction in emission compared to IC engine rickshaws only. This amount of CO_2 emission reduction does not make any considerable difference to support the utmost need to accomplish emission reduction targets to recover the quality of

Table 7 Well to tailpipe CO_2 emission in Mt/year for penetration of electric rickshaws in steps of 10% using coal energy-based power generation

Electric rickshaw penetration (%)	No. of electric rickshaws	CO_2 emission/year (million tonne/year) (EV)	No. of ICEV rickshaws	CO_2 emission/year (million tonne/year) (ICEV)	(EV + ICEV) Total CO_2 emission (million tonne/year)	Increase in CO_2 emission (million tonne/year)
0	0	0	7,500,000	12.37	12.37	0
10	750,000	1.78	6,750,000	11.13	12.91	0.54
20	1,500,000	3.56	6,000,000	9.89	13.45	1.08
30	2,250,000	5.33	5,250,000	8.66	13.99	1.62
40	3,000,000	7.11	4,500,000	7.42	14.53	2.16
50	3,750,000	8.89	3,750,000	6.18	15.07	2.71
60	4,500,000	10.67	3,000,000	4.95	15.61	3.25
70	5,250,000	12.44	2,250,000	3.71	16.15	3.79
80	6,000,000	14.22	1,500,000	2.47	16.70	4.33
90	6,750,000	16.00	750,000	1.24	17.24	4.87
100	7,500,000	17.78	0	0.00	17.78	5.41

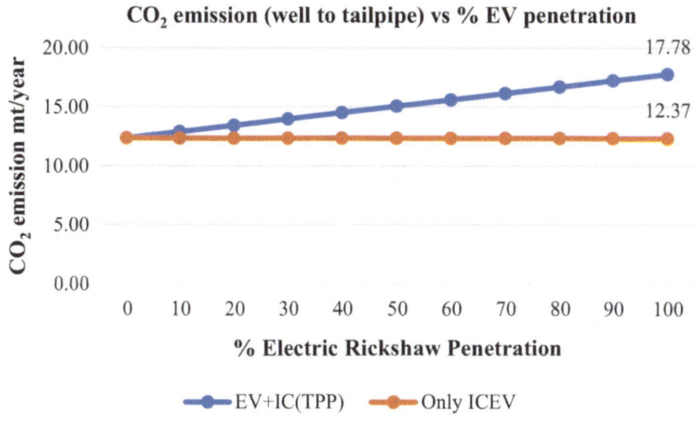

Fig. 7 Well to tailpipe CO_2 emission versus %EV rickshaws penetration

Table 8 Well to tailpipe CO_2 emission in Mt/year for penetration of electric rickshaws in steps of 10% using coal energy-based power generation with different plant efficiencies

Electric rickshaw penetration (%)	(EV + ICEV) Total CO_2 emission (million tonne/year) for different thermal power plant efficiency			
	25% eff	30% eff	35% eff	40% eff
0	12.37	12.37	12.37	12.37
10	12.91	12.61	12.40	12.24
20	13.45	12.86	12.43	12.11
30	13.99	13.10	12.47	11.99
40	14.53	13.35	12.50	11.86
50	15.07	13.59	12.53	11.74
60	15.61	13.84	12.57	11.61
70	16.15	14.08	12.60	11.49
80	16.70	14.32	12.63	11.36
90	17.24	14.57	12.67	11.24
100	17.78	14.81	12.70	11.11

air and lessen global warming. Therefore, switching to renewable energy sources for electricity generation is essential to achieve a sustainable and environmentally beneficial transition from internal combustion engines to electric vehicles [5].

The study by Evans S finds each kilowatt hour of electricity produced over the lifetime of a nuclear plant has an emissions footprint of 4 g of CO_2 equivalent (gCO_2e/kWh) and the footprint of solar comes in at 6 g CO_2eq/kWh and wind is also 4 g CO_2eq/kWh [38]. Considering these values, total CO_2 emission with solar and wind as source of energy to drive electric rickshaws is calculated, tabulated and plotted as shown in Table 9 and Fig. 9 respectively as given.

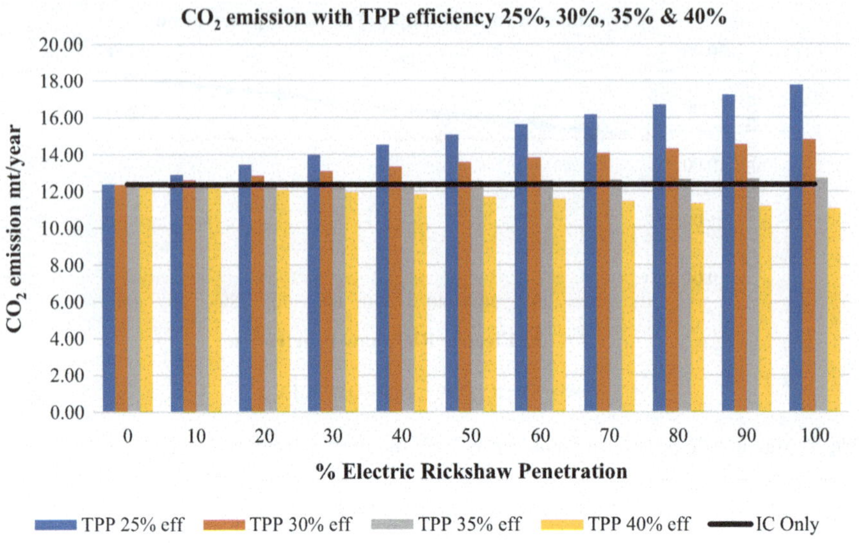

Fig. 8 Well to tailpipe CO_2 emission versus %EV rickshaws penetration with different TPP efficiencies

Table 9 CO_2 emission (million tonne/year) for penetration of electric rickshaws in step of 10% with solar and wind energy-based power generation

Electric rickshaw penetration (%)	No. of electric rickshaws	CO_2 emission/year by EV (million tonne/year)		Total CO_2 emission (EV + ICEV) (million tonne/year)		% Reduction in CO_2 emission (million tonne/year)	
		Solar power plant	Wind power plant	Solar power plant	Wind power plant	Solar power plant	Wind power plant
0	0	0	0	12.366	12.366	0.00	0.00
10	750,000	0.066	0.044	11.196	11.173	9.47	9.64
20	1,500,000	0.132	0.088	10.025	9.981	18.93	19.29
30	2,250,000	0.198	0.132	8.855	8.788	28.40	28.93
40	3,000,000	0.265	0.176	7.684	7.596	37.86	38.57
50	3,750,000	0.331	0.220	6.514	6.403	47.33	48.22
60	4,500,000	0.397	0.265	5.343	5.211	56.79	57.86
70	5,250,000	0.463	0.309	4.173	4.018	66.26	67.50
80	6,000,000	0.529	0.353	3.002	2.826	75.72	77.15
90	6,750,000	0.595	0.397	1.832	1.633	85.19	86.79
100	7,500,000	0.661	0.441	0.661	0.441	94.65	96.43

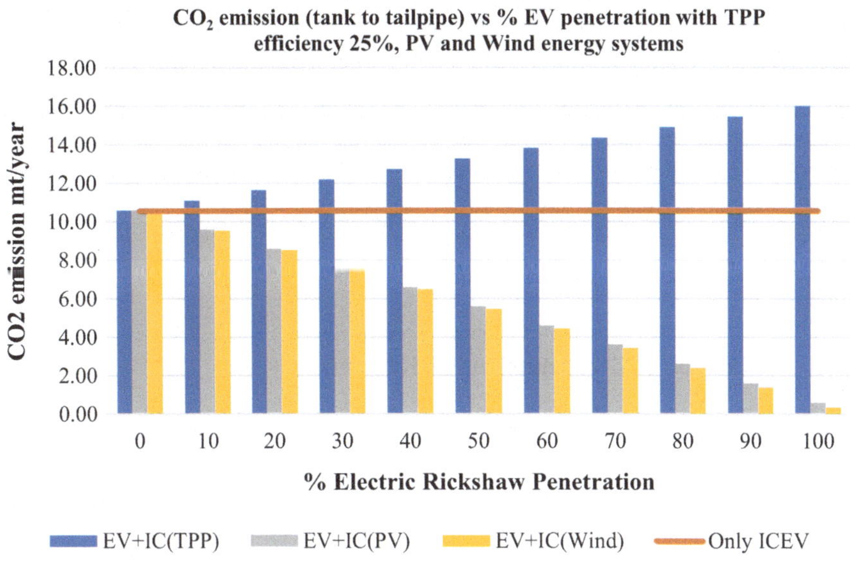

Fig. 9 Tank to tailpipe CO_2 emission (Mt/year) versus % electric rickshaws penetration

IC engine vehicle emissions are dispersing while EV emissions are localized (or concentrated). The IC engine emits gases directly to atmosphere as it travels. Such emissions cannot be contained. On the other hand, the emissions for electric vehicles (which is essentially electricity generated at power stations) are localized—in that they occur at a single location. Thus, these emissions can be captured and contained.

Also, in the analysis CO_2 emission from well to tailpipe is only considered. But Qiao et al. presented that the amount of CO_2 emission in manufacturing of electric vehicle is 60% more than conventional ICEV [25]. This is due to huge CO_2 emission that takes place in manufacturing of Li-Ion batteries, traction motors, electronic controllers, etc. This CO_2 emission during manufacture of vehicles can give a much better idea on viability of electric vehicles in Indian market.

4.3 *CO_2 Emissions for Lifetime Driving Distance*

In this analysis, we estimated the CO_2 emissions for lifetime distance travelled by Bajaj Compact RE rickshaw and Mahindra TREO with different thermal power plant efficiencies and also taking solar and wind energy being the primary energy resources. As per the report, the lifespan of autorickshaws is 15 years [31]. In the previous analysis of estimation of payback period as we had considered 300 working days with 70 kms travelling per day. So, the same data is considered to estimate the lifetime travelling distance and is found to be 315,000 kms, and for simplicity in calculation purpose, we rounded off the lifetime travelling distance to be 320,000 kms.

Table 10 CO_2 emission per kilometre for Mahindra TREO with different TPP efficiencies

Component	TPP—25%	TPP—30%	TPP—35%	TPP—40%
Energy required from coal (kWh)	36.395	30.329	25.996	22.747
Energy required from coal (kJ)	131,022.22	109,185.18	93,587.30	81,888.89
Mass of coal required (kg)	3.998	3.332	2.856	2.499
Burning 1 kg of coal emits (kg CO_2)	3.67	3.67	3.67	3.67
Total CO_2 emitted (kg CO_2)	14.674	12.228	10.481	9.171
CO_2 emission per kilometre (g/km)	112.873	94.061	80.624	70.546

4.3.1 Estimation of CO_2 Emission Per Kilometre

Following previously assumed/considered data is used to find out the CO_2 emission per kilometre for both rickshaws.

Bajaj Compact RE

- Distance travelled consuming 1 litre of petrol 35 kms
- 1 L of petrol emits 2.35 kg CO_2
- CO_2 emission due to 1 L of petrol production to user process 398 g/litre

So, the CO_2 emission per kilometre for Bajaj Compact RE is 78.51 g/km (Table 10).

Mahindra TREO

- Maximum distance travelled with full charged battery 130.00 km
- Capacity of the Li-Ion battery 7.37 kWh
- Efficiency of battery charger 90%
- Efficiency of electricity transmission 90%
- Actual energy drawn from TPP to charge the battery to full 9.10 kWh
- Efficiency of TPP 25, 30, 35 and 40%
- Calorific value of burning 1 kg of Coal 32,770.00 kJ/kg
- Life span of battery of Li-Ion battery 5 years and 80,000 kms
- Average battery emission factor for Li-Ion battery 177 kgCO_2/kWh [11]

So, manufacturing of a 7.37 kWh Li-Ion battery emits 1304 kg of CO_2. This value of emission adds up to lifetime distance travelled emissions every 80,000 kms when battery of Mahindra TREO is replaced. Considering these above data, CO_2 emissions

Table 11 Lifetime distance travelled CO_2 emission for Bajaj Compact RE and Mahindra TREO with different TPP efficiencies

	CO_2 emitted tonnes						
Kilometres	Bajaj Compact RE	Mahindra TREO 25% TPP eff	Mahindra TREO 30% TPP eff	Mahindra TREO 35% TPP eff	Mahindra TREO 40% TPP eff	Mahindra TREO—Solar energy	Mahindra TREO—Wind energy
0	0	1304	1304	1304	1304	1304	1304
20,000	1570	3561	3185	2916	2715	1312	1310
40,000	3141	4515	3762	3225	2822	1321	1315
60,000	4711	6772	5644	4837	4233	1329	1321
80,000	6281	10,334	8829	7754	6948	2642	2631
100,000	7851	11,287	9406	8062	7055	2650	2636
120,000	9422	13,545	11,287	9675	8466	2659	2642
140,000	10,992	15,802	13,169	11,287	9876	2667	2648
160,000	12,562	19,364	16,354	14,204	12,592	3980	3958
180,000	14,133	20,317	16,931	14,512	12,698	3989	3963
200,000	15,703	22,575	18,812	16,125	14,109	3997	3969
220,000	17,273	24,832	20,693	17,737	15,520	4005	3975
240,000	18,843	28,394	23,879	20,654	18,236	5318	5285
260,000	20,414	29,347	24,456	20,962	18,342	5327	5290
280,000	21,984	31,604	26,337	22,575	19,753	5335	5296
300,000	23,554	33,862	28,218	24,187	21,164	5343	5301

Note—Battery replaced at 80000, 160000 and 24000 kms

for lifetime distance travelled of 320,000 kms are estimated and plotted as shown in Table 11 and Fig. 10.

The first observation from the obtained results is that at zero kilometre, the electric rickshaws CO_2 emission is more than IC engine rickshaw by 1304 kg because of emissions due to Li-Ion battery manufacturing. This battery has to be replaced at every 80,000 kms, and in the intervals of 80000 kms, we can observe the spike in CO_2 emission until 240,000 kms. Results show that CO_2 emissions by electric rickshaw Mahindra TREO with thermal power plant efficiency of 25% and 30% will be emitting more CO_2 than IC engine rickshaw Bajaj Compact RE throughout the lifetime. As the thermal efficiency increases to 30%, Mahindra TREOs emissions are more or similar to that of Bajaj Compact RE at certain period of lifetime. But it never emits lesser CO_2 than that of Bajaj Compact RE. But, when plant efficiency is 40%, there are certain intervals in the early lifetime distance where CO_2 emissions from electric rickshaw tries to cross the IC engine rickshaw emission graph at 40,000, 70,000, 90,000 and 160,000 kms. This point is called as Distance of Intersection Point (DIP). After 160,000 kms, the CO_2 emission by electric rickshaw is comparatively lesser than IC engine rickshaw. And as expected, with solar and wind energy being

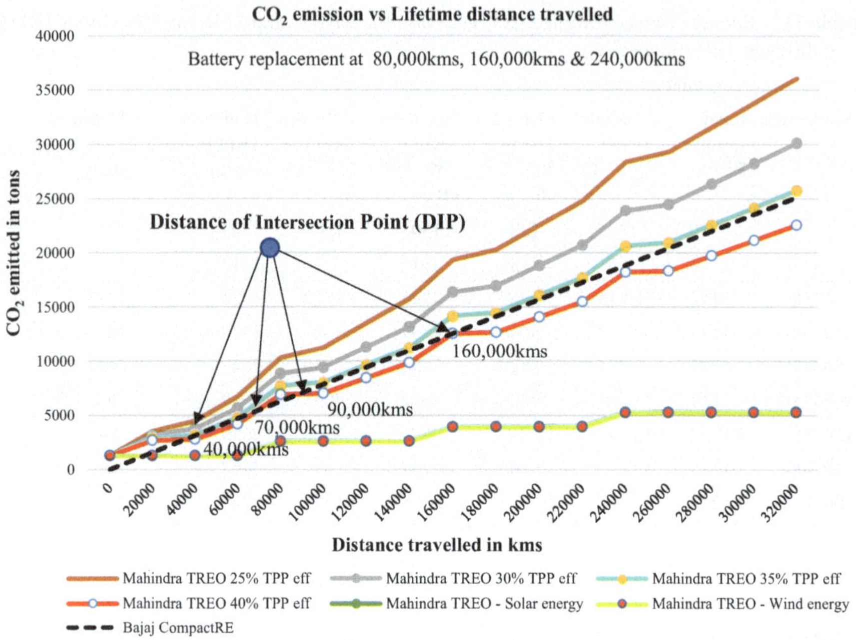

Fig. 10 Lifetime distance travelled CO_2 emission for Bajaj Compact RE and Mahindra TREO

the primary source of energy, the CO_2 emissions by electric rickshaw are much lower than that of IC engine rickshaw.

5 Conclusion

The study can be concluded as follows. In order to analyse the sustainability and implications viability of electric vehicle in Indian market, among the different segments three-wheeler auto rickshaw is considered. Cost analysis showed that, though the capital cost, maintenance cost and depreciation cost of electric rickshaw Mahindra TREO is higher than the IC engine rickshaw Bajaj RE, the lower fuel cost of Rs. 7980/- annually makes Mahindra TREO much economical with a net saving of Rs. 31,830/- annually after its estimated payback period of 2.84 years.

Secondly, GHG emission estimated for both Bajaj Compact RE and Mahindra TREO concluded that there is no significant amount of CO_2 emission reduction found. The analysis was carried out for both tank to tailpipe and well to tailpipe. Surprisingly, due to poor efficiency of thermal power plant assumed to be 25%, well to tailpipe CO_2 emission accountable by Mahindra TREO is approximately 43.76% more than conventional vehicle Bajaj Compact. The estimation was carried out by increasing the plant efficiency to 30%, 35% and 40%. At 40% TPP efficiency, the electric rickshaw CO_2 emissions were found to be less than IC engine rickshaw by 10.14% which is not sufficient enough to support the GHG emission reduction target set the Government of India. Then the further estimations made considering supply of electricity by renewable energy sources like solar and wind gave significant hopes on reduction in CO_2 emissions by 94% and 96% reduction respectively with 100% penetration of electric rickshaws in the market.

Thirdly, the effect of huge CO_2 emissions due to Li-Ion battery manufacturing was also considered to make the analysis more realistic. A lifetime travel distance of 320,000 kms is considered for a rickshaw and estimated the CO_2 emissions for different thermal power plant efficiencies and renewable energy sources as primary sources. Similar results were obtained showing that thermal plant efficiency of 40% would benefit in slightly lower GHG emissions. One solution to reduce the CO_2 emission impact on replacement of batteries is by recycling the batteries. Some studies have inferred that by recycling the materials the energy consumption and CO_2 emissions for manufacturing of new batteries almost gets reduced by 40–50%. But, it is challenging to develop a successful recycling system since batteries vary greatly in their chemical composition and structure. As a result, battery manufacturers prefer to use freshly mined metals over using recycled resources. Therefore, with this study we conclude that recycling of batteries and switching to renewable energy sources makes a sustainable and environmentally beneficial transition from internal combustion engines to electric vehicles.

References

1. IEA, *Data and Statistics—CO_2 Emissions by Sector.* Accessed 06 Sept 2022 (2019). https://www.iea.org/data-and-statistics
2. IEA, *India Energy Outlook.* International Energy Agency (2021). https://www.iea.org/reports/india-energy-outlook-2021
3. A. Jain, H.R. Jariwala, U. Mhaskar, Feasibility analysis for the penetration of electric vehicles in India in future, in *2019 International Conference on Computation of Power, Energy, Information and Communication (ICCPEIC)* (IEEE, Melmaruvathur, India, 2020), pp. 68–72. https://doi.org/10.1109/ICCPEIC45300.2019.9082380
4. NEMMP, *National Electric Mobility Mission Plan 2020 (NEMMP).* Ministry of Heavy Industries and Public Enterprises Government of India (2020). https://heavyindustries.gov.in/writereaddata/Content/NEMMP2020.pdf
5. V. Nimesh, D. Sharma, V.M. Reddy, A.K. Goswami, Implication viability assessment of shift to electric vehicles for present power generation scenario of India. Energy (Elsevier) **195**(116976) (2020). https://doi.org/10.1016/j.energy.2020.116976
6. Ministery of Power. 2021. *Power Sector at a Glance ALL INDIA.* https://powermin.gov.in/en/content/power-sector-glance-all-india.
7. G. Kalghatgi, Is it really the end of internal combustion engines and petroleum in transport? Appl. Energy (Elsevier) **225**, 965–974 (2018). https://doi.org/10.1016/j.apenergy.2018.05.076
8. A.F.N. Abdul, V.G. Manan, A.K. Zavaleta, G.K. Agarwal, A.A. Amer, Electrifying passenger road transport in India requires near-term electricity grid decarbonisation. Nat. Commun. **13**(2095), 1–13 (2022). https://doi.org/10.1038/s41467-022-29620-x
9. e-AMRIT, *Benefits of Electric Vehicles.* Edited by Government of India Niti Ayog. Accessed 13 June 2022 (2022). https://e-amrit.niti.gov.in/benefits-of-electric-vehicles

10. H. Xiong, H. Liu, R. Zhang, Y. Limin, Z. Zong, M. Zhang, Z. Li, An energy matching method for battery electric vehicle and hydrogen fuel cell vehicle based on source energy consumption rate. Int. J. Hydr. Energy (Elsevier) **44**(56), 29733–29742 (2019). https://doi.org/10.1016/j.ijhydene.2019.02.169
11. R. Kawamoto, H. Mochizuki, Y. Moriguchi, T. Nakano, M. Motohashi, Y. Sakai, A. Inaba, Estimation of CO_2 emissions of internal combustion engine vehicle and battery electric vehicle using LCA. Sustainability (MDPI) **11**(9) (2019). https://doi.org/10.3390/su11092690
12. E. Furuholt, Life cycle assessment of gasoline and diesel. Resour. Conserv. Recycl. (Elsevier) **14**(3–4), 251–263 (1995). https://doi.org/10.1016/0921-3449(95)00020-J
13. D. Kennedy, S.P. Philbin, Techno-economic analysis of the adoption of electric vehicles. Front. Eng. Manage. (Springer) **6**(4), 538–550 (2019). https://doi.org/10.1007/s42524-019-0048-x
14. S. Dang, A. Odonde, T. Mirza, C. Dissanayake, R. Burns, Sustainable energy management: an analysis report of the impacts of electric vehicles, in *14th International Conference on Environment and Electrical Engineering* (IEEE, Krakow, Poland, 2014), pp. 318–322. https://doi.org/10.1109/EEEIC.2014.6835886
15. T. Hara, Comprehensive comparison between hybrid and electric vehicles using a multi-dimensional techno-economic assessment diagram: concept and application. World Electric Veh. J. (MDPI) **9**(3–41) (2018). https://doi.org/10.3390/wevj9030041
16. G. Doluweera, F. Hahn, J. Bergerson, M. Pruckner, A scenario-based study on the impacts of electric vehicles on energy consumption and sustainability in Alberta. Appl. Energy **268**(114961) (2020). https://doi.org/10.1016/j.apenergy.2020.114961
17. S. Pareek, A. Sujil, S. Ratra, R. Kuma, Electric vehicle charging station challenges and opportunities: a future perspective, in *International Conference on Emerging Trends in Communication, Control and Computing (ICONC3)* (IEEE, Lakshmangarh, India, 2020). https://doi.org/10.1109/ICONC345789.2020.9117473
18. E.A.M. Falcão, A.C.R. Teixeira, J.R. Sodré, Analysis of CO_2 emissions and techno-economic feasibility of an electric commercial vehicle. Appl. Energy **193**, 297–307 (2017). https://doi.org/10.1016/j.apenergy.2017.02.050
19. J.S. Neubauer, A. Pesaran, B. Williams, M. Ferry, J. Eyer, A techno-economic analysis of PEV battery second use: repurposed-battery selling price and commercial and industrial end-user value. SAE Int. **01**(0349) (2012). https://doi.org/10.4271/2012-01-0349
20. T. Kobashi, M. Yarime, Techno-economic assessment of the residential photovoltaic systems integrated with electric vehicles: a case study of Japan. Energy Procedia (Elsevier) **158**, 3802–3807 (2019). https://doi.org/10.1016/j.egypro.2019.01.873
21. T. Kobashi, K. Say, J. Wang, M. Yarime, D. Wang, T. Yoshida, Y. Yamagata, Techno-economic assessment of photovoltaics plus electric vehicles towards household sector decarbonization in Kyoto and Shenzhen by the year 2030. J. Clean. Prod. (Elsevier) **253** (2020). https://doi.org/10.1016/j.jclepro.2019.119933
22. H.S. Hayajneh, S. Bashetty, M.N.B. Salim, X. Zhang, Techno-economic analysis of a battery energy storage system with combined stationary and mobile applications, in *IEEE Conference on Technologies for Sustainability (SusTech)* (IEEE, Long Beach, CA, USA, 2018), pp 1–6. https://doi.org/10.1109/SusTech.2018.8671332
23. A.M. Schetinger, D.H.N. Dias, B.S.M.C. Borba, G.D.P. da Silva, Techno-economic feasibility study on electric vehicle and renewable energy integration: a case study. Energy Storage (Wiley) **197** (2020). https://doi.org/10.1002/est2.197
24. K. Laurischkat, D. Jandt, Techno-economic analysis of sustainable mobility and energy solutions consisting of electric vehicles, photovoltaic systems and battery storages. J. Clean. Prod. (Elsevier) **179**, 642–661 (2018). https://doi.org/10.1016/j.jclepro.2017.11.201
25. Q. Qiao, F. Zhao, Z. Liu, S. Jiang, H. Hao, Comparative study on life cycle CO_2 emissions from the production of electric and conventional vehicles in China. Energy Procedia (Elsevier) **105**, 3584–3595 (2017). https://doi.org/10.1016/j.egypro.2017.03.827
26. S. Rohr, S. Wagner, M. Baumann, S. Müller, M. Lienkamp, A techno-economic analysis of end of life value chains for lithium-ion batteries from electric vehicles, in *Twelfth International Conference on Ecological Vehicles and Renewable Energies (EVER)* (IEEE, Monte Carlo, Manaco, 2017), pp. 1–14. https://doi.org/10.1109/EVER.2017.7935867

27. A.E.P. Abas, J. Yong, T.M.I. Mahlia, M.A. Hannan, Techno-economic analysis and environmental impact of electric vehicle. IEEE Access **7**, 98565–98578 (2019). https://doi.org/10.1109/ACCESS.2019.2929530
28. I. Morse, *A Dead Battery Dilemma*. Science.org. May 20. Accessed May 2021 (2021). https://www.science.org/content/article/millions-electric-cars-are-coming-what-happens-all-dead-batteries
29. P. Thakur, S. Pal, Estimating vehicular emissions from auto rickshaws plying in Bengaluru. Int. J. Scien. Eng. Res. (IJSER) **9**(5), 2241–2245 (2018). https://www.ijser.org/research-paper-publishing-may-2018_page5.aspx
30. TERI. *Road map for Electrification of Urban Freight in India*. The Energy and Resources Institute, New Delhi (2020). https://www.teriin.org/sites/default/files/files/Decarbonization_of_Transport%20Sector_in_India.pdf.
31. Hakeem Committee, *Report of the Committee for Determination of Fare Structures of Taxis and Auto rickshaw in Maharashtra State*. Government of Maharashtra (2017)
32. SIAM. *SIAM Annual Reports*. Society of Indian Automobile Manufacturers (2020). https://www.siam.in/cpage.aspx?mpgid=42&pgidtrail=89.
33. Bajaj, *Bajaj Compact RE Specifications*. 20 Feb (2021). https://trucks.cardekho.com/en/trucks/bajaj/compact-4s/specifications
34. Mahindra TREO. *mahindraelectric*. Accessed 20 Feb 2021 (2021). https://www.mahindraelectric.com/vehicles/treo-electric-auto/
35. MESCOM, *Electricity Tariff 2021*. Mangalore (2021). https://mescom.karnataka.gov.in/info-4/Electricity+Tariff+Schedule/en
36. SIAM. *Automobile Domestic Sales Trends*. Accessed 20 Feb 2021 (2021). https://www.siam.in/statistics.aspx?mpgid=8&pgidtrail=14.
37. T. Zhang, Methods of improving the efficiency of thermal power plants. J. Phys.: Conf. Ser. ISPECE 2019. IOP Publishing (2020). https://doi.org/10.1088/1742-6596/1449/1/012001
38. S. Evans, *Solar, wind and nuclear have 'amazingly low' carbon footprints, study finds* (2017). https://www.carbonbrief.org/solar-wind-nuclear-amazingly-low-carbon-footprints

A Numerical Study on a c-Si(P) Substrate-Based Homo-Hetero Junction Solar Cell

Himangshu Deka, Arun Kumar Sunaniya, and Pratima Agarwal

Abstract A homo-hetero junction solar cell made up of crystalline and amorphous silicon is investigated using AFORS-HET simulator. The ability of a homo-hetero junction cell to improve photoelectric behavior is well established. The study is carried out on a p-type crystalline silicon substrate which forms a shallow junction with n-type crystalline silicon. The addition of highly doped amorphous silicon at the top and bottom surfaces creates extra field, thus reducing the interface recombination and series resistances. The study includes the thickness variation of n-type amorphous silicon, n-type crystalline silicon, and p-type substrate, to find out the optimum result. Also, the effect on cell performances due to the variation of n-type amorphous silicon doping is performed. In comparison with a heterojunction with thin intrinsic layer (HIT) cell, homo-hetero junction is different in a way that it includes the additional crystalline silicon layer which actually enhances the field-effect passivation, thereby further reducing the series resistance and enhancing the fill factor to a greater extent. The excess amount of field also makes it less sensitive toward the interfacial defect states, and to prove it, we increase the defect states up to 10^2 times greater than that of a HIT cell. The calculated total recombination factor, J_o, is found to be as low as 3.45 fA/cm^2. Finally, we achieved an efficiency of 28.33% with enhanced fill factor (FF) of 0.88.

Keywords AFORS-HET · Amorphous silicon · Crystalline silicon · HIT

1 Introduction

Despite the availability of a different materials for fabricating a heterojunction solar cell, the industry currently favors silicon-based crystalline/amorphous heterojunction

H. Deka · A. K. Sunaniya (✉)
Department of Electronics and Instrumentation Engineering, National Institute of Technology Silchar, Silchar, Assam, India
e-mail: arun@ei.nits.ac.in

P. Agarwal
Department of Physics, Indian Institute of Technology, Guwahati, Assam, India

V. S. Moholkar et al. (eds.), *Sustainable Energy Generation and Storage*,
https://doi.org/10.1007/978-981-99-2088-4_7

solar cells. Regardless of other materials' popularity among researchers, monocrystalline, microcrystalline, and amorphous-based silicon remain the common utilized variants of silicon for the fabrication of solar devices, accounting for nearly 90% of solar technology globally [1]. The belief in silicon stems from its bandgap appropriateness for photovoltaics and its global availability. A thin intrinsic type amorphous hydrogenated silicon (i-a-Si:H) with better surface passivating nature is commonly put in between the crystalline silicon (c-Si) base and a highly doped amorphous type silicon [2] commonly known as HIT cell. The company called Sanyo, now Panasonic, has designed this type of cells with a conversion efficiency of up to 24.7 percent. Several well-known groups are also working on these types of cells [3–5]. Even though i-a-Si:H layer passivation's effect, inclusion of an inherent thin a-Si has adverse impacts because of its high resistivity, which in turn reduces the fill factor (FF) [6, 7]. As a result, it is critical to create a new type of passivating layer to substitute the said layer that can improve FF although maintaining a similar passivation effect. The N/N++ and P/P++ type homo-type junctions are well known for being employed as a field-effect passivating layers to minimize recombination rates at back [8], again high efficient homojunction-type solar devices use the N/N++ or P/P++ type junction to create a passivation effect at the front surfaces [9]. Harder [10] suggested a new cell structure that combines both homo- and heterojunctions (homo-hetero junctions) to minimize hetero interface recombination of a solar cell. Later, Zhong et al. [11] have demonstrated the properties and physical aspect of n-type wafer-based P+-a-Si/P-c-Si/N-c-Si/N+-a-Si solar cell. Zhong et al. have claimed to have better FF than the HIT cell, and also, the negative effect of interface defect on cell performances is lowered. In this study, we have investigated a n+-a-Si:H/c-Si(n)/c-Si(p)/p+-a-Si:H type homo-hetero junction cell on a p-type wafer. Despite the benefits of n-type silicon, the p-type silicon dominates the photovoltaic market, however only for homojunction-type solar cells. Because the p-type doped silicon solar cells have lower irradiation degradation than n-type doped silicon-based cells, they are recommended for space applications over n-type doped silicon [12]. In the p-type doped crystalline silicon, electrons, the minority carrier, have a greater diffusion length and greater mobility values than holes. Apart from that, p-type doped wafers are less expensive and widely used in the micro-electronics industry; thus, p-type silicon wafers are always preferable. Another difference with Zhong et al. reported structure is that here the doping of crystalline silicons is kept minimum so that the damage due to excessive doping does not take place in the actual wafer. The heavily doped n+-a-Si:H with respect to c-Si (n) acts as a front surface field as well as it controls the band bending behavior of PN homojunction. The extra field that is generated at the junction of c-Si (n) and the n+-a-Si:H layer reduces the interface recombination, lowers the series resistance, and hence enhances the FF. Our study includes the thickness variation of p-type substrate, n-type doped crystalline silicon layer, and n-type doped amorphous silicon to find out the optimum result. The impact of n-type crystalline silicon doping concentration is also studied. Finally, with a V_{oc} of 0.763 V, a J_{sc} of 42.02 mA/cm^2, and FF of 0.88, we were able to attain a 28.33% efficiency.

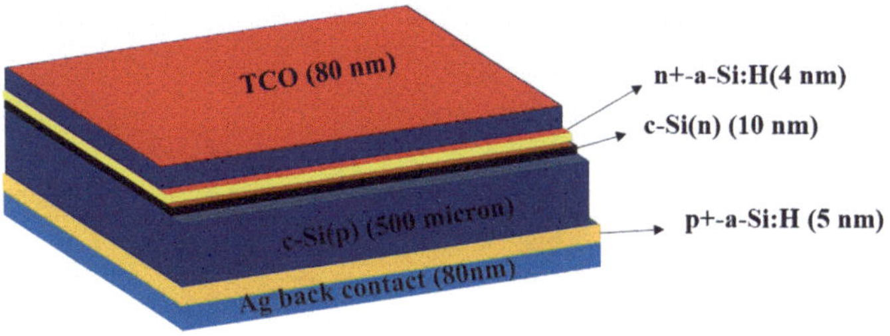

Fig. 1 Simulation structure

2 Simulation Model with Parameter Details

In this study, a simulator namely AFORS-HET is used to study the n+-a-Si:H/c-Si(n)/c-Si(p)/p+-a.-Si:H type homo-hetero-type junction solar cell structure shown in Fig. 1. This is a one-dimensional numerical software that readily handles multiple layered device and can be used for homo- and heterojunction models. For the numerical part, the simulator solves Poisson's equation and the Continuity equation at the same time. However, for the sake of simplicity, the analytical section has been considered under specific assumptions.

We used a p-type crystalline wafer carrier doping density of 5×10^{16} cm^{-3} in our simulation, which fits the findings of Xixing Wen et al. [12]. The crystalline silicon is chosen of good quality, having lone donor-type defect density of 10^{10} cm^{-3} at the bandgap's midpoint with electrons and holes cross section of 10^{-14} cm^2. The contact boundary at the front is assumed to be flat band; however, band bending is considered on the backside by setting Ag (silver) work function to 4.74 eV ((111) crystal orientation). Recombination velocities on both the front side and back side contact surfaces were set to 1×10^7 cm-s^{-1}. Table 1 displays the final simulation settings obtained from several references [13–15].

3 Results with Discussion

3.1 Thickness Optimization for Different Layer

The thickness optimization is done to get the best suitable thickness for the cell, and while doing this, we changed the thickness of each layer at a time and by fixing the thickness of the others as shown in Table 1. Figure 2a–c illustrate the corresponding results. The c-Si(p) layer is varied from 100 to 500 μm, c-Si(n) layer from 10 to 100 nm, and n-a-Si:H from 4 to 20 nm. During thickness variation, all other parameters are kept same as listed in Table 1. The cell performs best when the c-Si(p) layer is 500 μm thick, the c-Si(n) layer is of 10 nm thick, and the n-a-Si:H layer is 4 nm thick. As depicted in Fig. 2a, with increase in thickness, V_{oc} and FF decrease.

Table 1 Simulation parameter details

Numerical parameters	n+-a-Si:H	p+-a-Si:H	p-c-Si	n-c-Si
Layer's thickness	4×10^{-9}, (variable)	5×10^{-9}	500×10^{-6} (variable)	10×10^{-9} (variable)
Bandgap (eV)	1.72	1.72	1.17	1.17
Optical bandgap (eV)	1.72	1.72	1.17	1.17
Donor-type doping (cm^{-3})	5×10^{19}, (variable)	0	0	1.5×10^{15}
Acceptor-type doping (cm^{-3})	0	5×10^{19}	1.5×10^{16}	0
Dielectric constant of material	11.9	11.9	11.9	11.9
Electron affinity of material (eV)	3.9	3.9	4.05	4.05
Conduction band density (CB) (cm^{-3})	1×10^{20}	1×10^{20}	2.8×10^{19}	2.8×10^{19}
Valance band density (VB) (cm^{-3})	1×10^{20}	1×10^{20}	2.6×10^{19}	2.6×10^{19}
Mobility of electron (hole) (cm^2/Vs)	20 (5)	20 (5)	1107 (424)	1107 (424)
Net trap density in C B (cm^{-3})	1.36×10^{19}	1.36×10^{19}	–	–
Net trap density in V B (cm^{-3})	1.88×10^{19}	1.88×10^{19}	–	–
Urbach energy of CB tail (VB tail) (eV)	0.037 (0.081)	0.037 (0.081)	–	
Maximum Acceptor (A)-like Gaussian density of states (cm^{-3}/eV)	1.99×10^{19}	1.99×10^{19}	–	–
Maximum Donor (D)-like Gaussian state density (cm^{-3}/eV)	1.99×10^{19}	1.99×10^{19}	–	–
	7×10^{-16} (7×10^{-16})	7×10^{-16} (7×10^{-16})	–	–
Thermal electron (hole) cross section for VB tail (cm^2)	7×10^{-16} (7×10^{-16})	7×10^{-16} (7×10^{-16})	–	–
Standard deviation for donor (eV)	0.2	0.2	–	–

(continued)

Table 1 (continued)

Numerical parameters	n+-a-Si:H	p+-a-Si:H	p-c-Si	n-c-Si
Standard deviation for acceptor (eV)	0.2	0.2	–	–
Thermal electron (hole) cross section for acceptor like Gaussian state (cm^2)	3×10^{-15} (3×10^{-14})	3×10^{-15} (3×10^{-14})		
Thermal electron (hole) cross section for donor-like Gaussian state (cm^2)	3×10^{-14} (3×10^{-15})	3×10^{-14} (3×10^{-15})		

The decreases in V_{oc} are due to the recombination effect as thickness increases and FF is reduced because of series resistance that experienced due to thicker absorber layer. The thick absorber layer, which can absorb longer wavelengths photon, may explain the higher J_{sc}. Because a high-quality crystalline absorber layer having a longer diffusion length was chosen, the electron–hole pairs that generated even at a deeper depth of the absorber layer now can effortlessly reach the junction. The phenomenon can be well understood by the external quantum efficiency (E.Q.E) curve discussed later. As the J_{sc} increases, hence the efficiency of the cell is also improved with thicker absorber layer.

From Fig. 2b, it can be noted that with the increment in thickness of the c-Si(n), V_{oc} and FF remain almost unchanged; however, a decreased trend can be seen in case of J_{sc}, hence the efficiency. Again in Fig. 1c, with increase in thickness of n-a-Si:H layer, the V_{oc} and J_{sc}, are seen to be decreasing which ultimately decreases the efficiency. The decrease in J_{sc} in Fig. 1b, c may be due to the reason that some of the generated electron–hole pairs may have to travel a bit more than their diffusion length with increase in thickness. This leads to the enhancement of recombination, hence affecting the V_{oc} in Fig. 1c.

3.2 *Doping Optimization of n-a-Si:H Layer*

The variation in carrier doping is established within a variation from 5×10^{-15} to 5×10^{-19} cm^{-3}. The motive is to keep the carrier concentration of n-a-Si:H greater than that of the c-Si(n) layer so that n-a-Si:H behaves like a heavily doped layer in comparison with c-Si(n) layer and thus establish a high front surface field at the metallurgical junction. Figure 3 depicted the outcomes of carrier doping on various parameters. As the doping increases, a junction forms at the meeting point of n-a-Si:H and c-Si(n) and hence the electric field. As the n-a-Si:H layer doping increases, the shifting of depletion layer width between n-a-Si:H and c-Si (n) is more toward

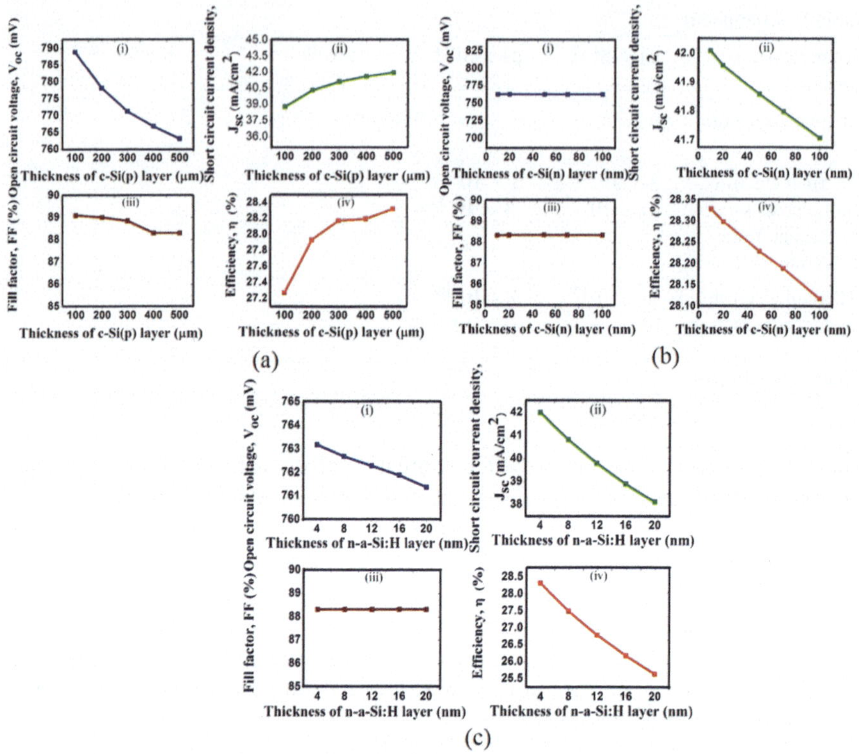

Fig. 2 Layer thickness variation of (**a**) c-Si(p), (**b**) c-Si(n), and (**c**) n-a-Si:H

c-Si(n) layer; hence, the field at x_p (edge of depletion layer width at c-Si(n) side) has complete control on the charge carrier recombination at surface as well as on the band bending of homojunction between c-Si(n) and c-Si(p). The lower recombination at the surface causes to increase the V_{oc} and with the increase in n-a-Si:H layer doping a downhill trend for electrons is formed; thus, it becomes easier for electron to reach the contact causing increase in the J_{sc} and hence the efficiency.

3.3 J–V *and EQE Responses*

The *J–V* response was measured under the usual AM 1.5 sun spectrum as shown in Fig. 4. The measured parameters from the *J–V* response are $V_{oc} = 0.763$ V, $J_{sc} = 42.02$, FF $= 0.88$, and $\eta = 28.33\%$. The lower V_{oc} is because of the reduced total recombination factor. The total recombination factor is measured as low as 3.45 fA/cm^2 which is far lower than our previously reported model [15] of 16.1 fA.cm^{-2}. Low surface recombination could explain the higher J_{sc}. Most importantly, the *FF* has been improved a lot in comparison with HIT solar cell which was one of the

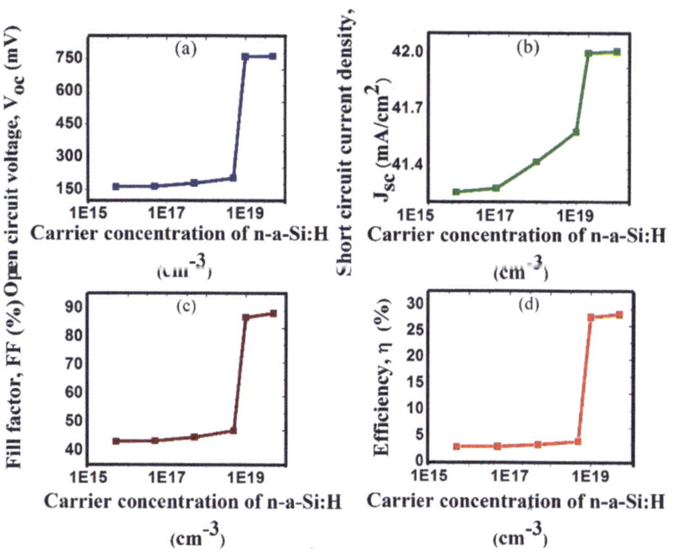

Fig. 3 Carrier concentration variation of n-a-Si:H layer

prime aims of this model. The improvement in FF is due to the additional field which reduces the series resistance. Initially, the performances were measured at the interface density of defect states of 1×10^{10} cm^{-2}. To check the passivation effect of the cell, we increased the density of defect states up to 1×10^{12} cm^{-2}. However, the results were unaltered; hence, we consider the higher density of defect states in this study.

The E.Q.E responses were taken under the illumination setting of 1×10^{16} cm^{-2}.s^{-1} monochromatic intensity. Figure 5 shows the E.Q.E responses for the thickness variation of c-Si(p) layer for the range of 100–500 μm.

Fig. 4 *J–V* response of homo-hetero junction solar cell

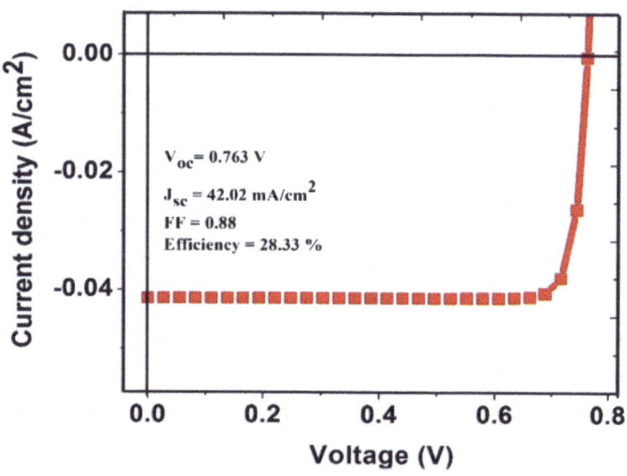

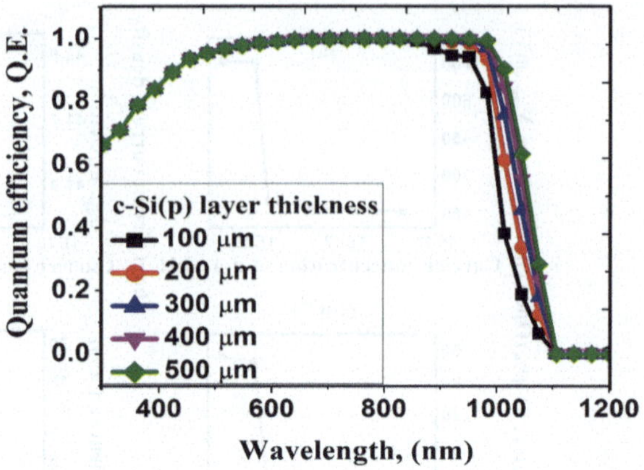

Fig. 5 E.Q.E responses for varying thickness of c-Si(p) layer

Table 2 Comparative results

Cell type	$V_{oc.}$ (V)	$J_{sc.}$ (mA/cm^2)	FF	Eff. (%)
p-a-Si:H/i-a.-Si:H/c-Si(n)/I-a-Si:H/n-a-Si:H (HIT) [11]	0.747	39.39	0.80	23.94
p+-a-Si:H/c-.Si(p)/c-Si(n)/n+-a-Si:H (Homo-hetero junction) [11]	0.750	40.36	0.84	25.37
n+-a-Si:H/c-Si(n)/c-Si(p)/p+-a-Si:H (Present study)	0.763	42.02	0.88	28.33

With the increment in thickness of c-Si(p) layer, a notable improvement in E.Q.E. responses can be observed specially in the wavelength range of 850–1100 nm indicating the absorption of longer wavelengths at higher depth.

The obtained results are finally compared with the previously reported results. All the parameters are seen to be improved than the previously reported results, and this is because of the change we have incorporated in the homo-hetero junction model (Table 2).

4 Conclusion

The simulation study of a homo-hetero junction with cell type n+-a-Si:H/c-Si(n)/c-Si(p)/p+-a-Si:H is studied using AFORS-HET simulator. The main objective of enhancing the FF than that of a HIT cell has been achieved with 0.88 FF. The inclusion of an extra c-Si(n) layer generates additional field which further helps in reducing the total recombination factor to a value of as low as 3.45 fA/cm^2. Considering some other studies like variation of layers thickness and carrier doping variation of the n-a-Si:H layer, finally the cell was optimized to deliver an efficiency of 28.33%.

References

1. S.S. Hegedus, A. Luque, Status, trends, challenges and the bright future of solar electricity from photovoltaics, in *Handbook of Photovoltaic Science and Engineering* (2003), pp. 1–43. https://doi.org/10.1002/0470014008.ch1
2. S.D. Wolf, Intrinsic and doped a-Si: H/c-Si interface passivation, in *Physics and Technology of Amorphous-Crystalline Heterostructure Silicon Solar Cells* (2012), pp. 223–259. https://doi.org/10.1007/978-3-642-22275-7_7
3. S. Olibet, E. Vallat-Sauvain, L. Fesquet, C. Monachon, A. Hessler-Wyser, J. Damon-Lacoste, S. De Wolf, C. Ballif, Properties of interfaces in amorphous/crystalline silicon heterojunctions. Physica Status Solidi (a) **207**(3), 651–656 (2010). https://doi.org/10.1002/pssa.200982845
4. S. Miyajima, J. Irikawa, A. Yamada, M. Konagai, High-quality nanocrystalline cubic silicon carbide emitter for crystalline silicon heterojunction solar cells. Appl. Phys. Lett. **97**(2), 023504 (2010). https://doi.org/10.1063/1.3460917
5. V. Kanneboina, R. Madaka, P. Agarwal, Stepwise tuning of the doping and thickness of a-Si: H (p) emitter layer to improve the performance of c-Si (n)/a-Si: H (p) heterojunction solar cells. J. Mater. Sci.: Mater. Electron. 1–9 (2021). https://doi.org/10.1007/s10854-020-05187-5
6. M. Rahmouni, A. Datta, P. Chatterjee, J. Damon-Lacoste, C. Ballif, P. Roca i Cabarrocas, Carrier transport and sensitivity issues in heterojunction with intrinsic thin layer solar cells on N-type crystalline silicon: a computer simulation study. J. Appl. Phys. **107**(5), 054521 (2010). https://doi.org/10.1063/1.3326945
7. G. Garcia-Belmonte, J. García-Cañadas, I. Mora-Seró, J. Bisquert, C. Voz, J. Puigdollers, R. Alcubilla, Effect of buffer layer on minority carrier lifetime and series resistance of bifacial heterojunction silicon solar cells analyzed by impedance spectroscopy. Thin Solid Films **514**(1–2), 254–257 (2006). https://doi.org/10.1016/j.tsf.2006.02.020
8. X. Gu, X. Yu, D. Yang, Efficiency improvement of crystalline silicon solar cells with a back-surface field produced by boron and aluminum co-doping. Scripta Mater. **66**(6), 394–397 (2012). https://doi.org/10.1016/j.scriptamat.2011.11.044
9. F. Granek, M. Hermle, D.M. Huljić, O. Schultz-Wittmann, S.W. Glunz, Enhanced lateral current transport via the front N+ diffused layer of n-type high-efficiency back-junction back-contact silicon solar cells. Prog. Photovoltaics Res. Appl. **17**(1), 47–56 (2009). https://doi.org/10.1002/pip.862
10. N.-P. Harder, Heterojunction solar cell with absorber having an integrated doping profile, Patent 017 437 4A1 (2011)
11. S. Zhong, X. Hua, W. Shen, Simulation of high-efficiency crystalline silicon solar cells with homo–hetero junctions. IEEE Trans. Electron Devices **60**(7), 2104–2110 (2013). https://doi.org/10.1109/TED.2013.2259830
12. J. Cotter, J. Guo, P. Cousins, M. Abbott, F. Chen, K. Fisher, P-type versus n-type silicon wafers: prospects for high-efficiency commercial silicon solar cells. IEEE Trans. Electron Devices **53**(8), 1893–1901 (2006). https://doi.org/10.1109/TED.2006.878026
13. F. Wang, Y. Gao, Z. Pang, L. Yang, J. Yang, Insights into the role of the interface defects density and the bandgap of the back surface field for efficient p-type silicon heterojunction solar cells. RSC Adv. **7**(43), 26776–26782 (2017). https://doi.org/10.1039/C7RA04018K
14. X. Hua, Z. Li, W. Shen, G. Xiong, X. Wang, L. Zhang, Mechanism of trapping effect in heterojunction with intrinsic thin-layer solar cells: effect of density of defect states. IEEE Trans. Electron Devices **59**(5), 1227–1235 (2012). https://doi.org/10.1109/TED.2012.2186139
15. H. Deka, A.K. Sunaniya, P. Agarwal, Design and simulation of highly efficient one-sided short PIN diode silicon heterojunction solar cell. IEEE J. Photovoltaics **12**(1), 204–212 (2021). https://doi.org/10.1109/JPHOTOV.2021.3116016

Optimization and Simulation of Bifacial Heterojunction Solar Cell with Gradient Doping Using AFORS-HET

Gaurav Singh and Pratima Agarwal

Abstract Fabrication of solar cells is an expensive and resource-consuming process due to which we cannot use hit and trial methods to obtain the best output results. This simulation work on Heterojunction Bifacial Solar Cell can provide the optimum values of various parameters that affect the output efficiency of the solar cell. Therefore, the cost of redundant fabrication of solar cells can be eliminated. In this study, we have simulated a Bifacial Heterojunction Solar Cell (TCO/a-Si:H(p)/a-Si:H(i)/c-Si(n)/a-Si:H(i)/a-Si:H(n+)/TCO/Ag). The parameter variations such as emitter layer thickness, uniform vs gradient doping of emitter layer and the back surface field (BSF) layer, and the introduction of texture in the front transparent conducting oxide (TCO) layer were studied and their effect on open-circuit voltage (V_{oc}), short circuit current density (J_{sc}), fill factor (%FF) and output efficiency (%η) were observed. The best outcome with an output efficiency of 16.37% is obtained with the fill factor of 81.93%. At this efficiency, the open-circuit voltage (V_{oc}) of 613.7 mV and short circuit current density (J_{sc}) of 32.56 A/cm^2 are observed.

Keywords HIT—Heterojunction with intrinsic thin layer · TCO—Transparent conducting oxide · PV—Photovoltaic · PECVD—Plasma-enhanced chemical vapour deposition · AM—Air mass · BSF—Back surface field · Ag—Silver

1 Introduction

The energy need of the ever-growing population is the burgeoning issue that needs to be addressed in the present time, and the whole world is shifting towards harvesting renewable energy. 173 countries around the world have set their targets for renewable energy with 146 countries having support policies for it [1]. Out of many forms of

G. Singh (✉) · P. Agarwal
School of Energy Science and Engineering, Indian Institute of Technology Guwahati, Guwahati, Assam 781039, India
e-mail: gauravsingh@iitg.ac.in

P. Agarwal
Department of Physics, Indian Institute of Technology Guwahati, Guwahati, Assam 781039, India

V. S. Moholkar et al. (eds.), *Sustainable Energy Generation and Storage*,
https://doi.org/10.1007/978-981-99-2088-4_8

renewable energy sources, the most prominent ones are wind energy and solar energy. India has set a target to fulfill its 40% energy demand with renewable energy sources by the year 2030 [2]. On 11 January 2010, India took a huge initiative by launching Jawaharlal Nehru National Solar Mission to make India a global leader in solar energy by installing 100 GW of solar PV by 2022 [3].

The yearly incident solar energy over India is 5000 trillion kWh with an average of 4–7 kWh per sq.m per day providing a huge prospective for the solar PV market in India [4]. The output generated by solar PV depends upon many factors like the type of PV material, solar radiation, module orientation, weather, cell temperature, geographical location, dust, inverter efficiency, etc.

Silicon solar cells with a heterojunction structure of a-Si:H/c-Si are considered one of the highest efficiency solar cells. The performance of heterojunction solar cells depends upon the interface quality and transport properties of the charge carriers [5]. Interface quality determines the efficiency of a solar cell, and hence, most of the studies are focused on its optimization. In 1994, 20% efficiency was achieved for the solar cell of 1 cm^2 aperture by SANYO with an intrinsic thin layer of a-Si:H between p-type a-Si:H/n-type c-Si using the process known as plasma-enhanced chemical vapour deposition (PECVD) [6]. This structure has come to be known as the P-I-N solar cell structure.

Bifacial Solar Cells can be the next advancement in the field of solar photovoltaic. The first Bifacial Solar Cell was proposed by the Japanese researcher 'mori' in 1960. This p^+np^+ solar cell structure has a collecting pn junction on both surfaces of the silicon wafer. The main objective of this structure was to increase the conversion efficiency of silicon solar cells which was limited due to the minority carriers at that time. Because of the pn junction at the rear side of the solar cell, the collection efficiency for long-wavelength photons was improved. Mori's structure enabled the possibility of providing double-sided illumination to the solar cells by the means of mirrors.

The first HIT bifacial solar cells were also made by inserting a thin intrinsic layer of a-Si:H between n-type c-Si and the back surface a-Si:H layer. Due to this, the rear surface recombination velocity was reduced resulting in an increase in efficiency beyond 20% [6].

2 Model

2.1 Structure

The simulation of solar cell structure for this study is performed with the Afors-Het simulation tool. This software provides the numerical solution to the one-dimensional equations of the semiconductor structure with required boundary conditions under Direct Current (DC) conditions. For our study, we have defined our solar cell structure with TCO/a-Si:H(p)/a-Si:H(i)/c-Si(n)/a-Si:H(i)/a-Si:H(n+)/TCO/Ag configuration as shown in Figs. 1 and 2.

From the two figures, it is clear that we have defined two separate structures for analyzing the effect of the textured TCO layer on the output parameters of the solar cells since the textured layer absorbs more light with less reflection and hence increasing the excess of photons to the further subsequent layers inside the solar cell. The second layer of the structure is the P-type hydrogenated amorphous silicon layer also known as the emitter layer. The third layer of the structure is the intrinsic thin hydrogenated amorphous silicon layer. The fourth layer of the structure is the N-type crystalline silicon layer also known as the base layer. The fifth layer of the structure is the intrinsic thin hydrogenated amorphous silicon layer. The sixth layer

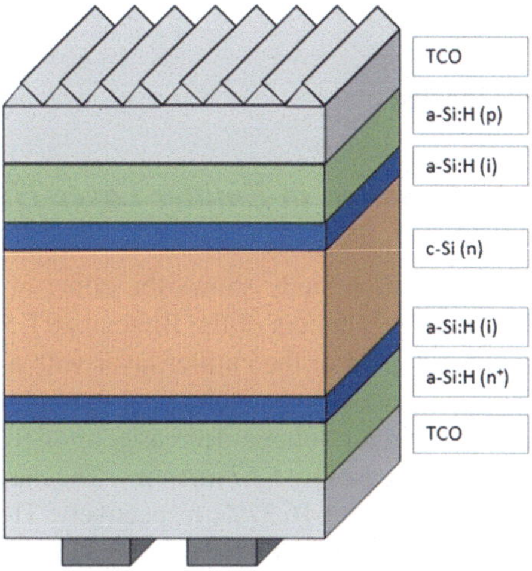

Fig. 1 Bifacial structure of HIT solar cell with surface texture

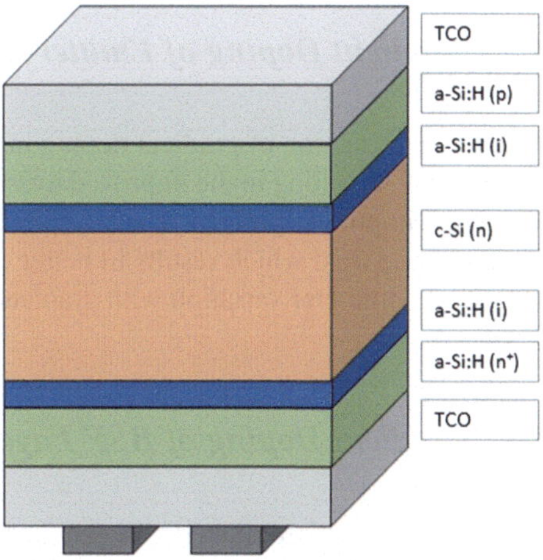

Fig. 2 Bifacial structure of HIT solar cell without surface texture

is the back absorption layer of N-type hydrogenated amorphous silicon, and the last layer is the back TCO layer with silver metal grid contacts. The solar illumination with Air Mass (AM) 1.5 for the power density of 1000 W/m^2 and wavelength range of 300–1100 nm is adopted for this study. The reflectivity of 0.1 and 1 was set for the front and back TCO layers, respectively. For this study, we have not introduced any defect states.

2.2 Simulation Parameters of Bifacial Solar Cell Layers

See Table 1.

3 Results

3.1 Variation in Emitter Layer Thickness

This simulation study shows the effect of variation of emitter layer thickness on the output parameters of the Bifacial HIT Solar Cell. In this study, we have chosen gradient doping of the emitter layer with a textured TCO layer. From Table 2, it is observed that, with the increase in the thickness of the emitter layer from 3 to 9 nm, the open-circuit voltage decreases up to 1.7 mV whereas the short circuit current density decreases by 1.67 mA/cm^2. The maximum fill factor and efficiency observed were 81.99% and 16.37%, respectively. The best performance of the solar cell was obtained at 3 nm thickness with 16.37% efficiency.

3.2 Gradient Doping of Emitter Layer

With the introduction of gradient doping in the front TCO layer, the output parameters increase, resulting in the improved overall output efficiency of the solar cell. The reason for improved performance with gradient doping is the introduction of additional electric field which results in better charge carrier separation. Table 3 shows the output parameter variation with gradient doping of the emitter layer.

3.3 Gradient Doping of BSF Layer

The introduction of gradient doping on the back surface field layer (BSF) has shown no effect on the output parameters. Table 4 shows the results obtained from this study.

Table 1 Different layer parameters used for the simulation [7]

Sl. No.	Parameter	Emitter a-Si:H (p)	Intrinsic a-Si:H (i)	Base c-Si (n)	BSF a-Si:H (p)
1	Layer thickness (nm)	3–9	1	3×10^5	5
2	Dielectric constant	11.9	11.9	11.9	11.9
3	Electron affinity (eV)	3.75	3.75	4.05	3.75
4	Mobility band gap (eV)	1.73	1.73	1.12	1.73
5	Optical band gap (eV)	1.73	1.73	1.12	1.73
6	Effective DOS in conduction band Nc (cm^{-3})	1×10^{20}	1×10^{20}	2.84×10^{19}	1×10^{20}
7	Effective DOS in valence band Nv (cm^{-3})	1×10^{20}	1×10^{20}	1.04×10^{19}	1×10^{20}
8	Mobility of electrons (cm^2 V^{-1} S^{-1})	5	20	1040	10
9	Mobility of holes (cm^2 V^{-1} S^{-1})	1	2	412	1
10	Acceptor doping (cm^{-3})	1×10^{19} to 2×10^{20}	0	0	0
11	Donor doping (cm^{-3})	0	0	5×10^{16}	1×10^{19} to 2×10^{20}
12	Thermal velocity of electrons (cm S^{-3})	1×10^7	1×10^7	1×10^7	1×10^7
13	Thermal velocity of holes (cm S^{-3})	1×10^7	1×10^7	1×10^7	1×10^7
14	Density of layers (g cm^{-3})	2.328	2.328	2.328	2.328

3.4 Effect of Texture on TCO Layer

Figures 1 and 2 show the structure of solar cell with and without textured TCO layer, respectively. With the introduction of the texture into the front TCO layer, the effective surface area of the solar cell increases. This increased surface area allows more sun illumination to fall on it resulting in more photon absorption. Furthermore,

Table 2 Effect of variation of emitter layer thickness with gradient doping (from 2e20 to 1e19) and textured TCO

S. No.	Emitter thickness (nm)	V_{OC} (mV)	J_{SC} (mA/cm^2)	FF (%)	Efficiency (%)
1	3	613.7	32.56	81.93	16.37
2	5	613.7	31.97	81.84	16.06
3	7	612	31.41	81.99	15.76
4	9	612	30.89	81.91	15.48

Table 3 Gradient doping versus uniform doping of the emitter layer with a thickness of 3 nm and textured TCO

S. No.	Emitter thickness (nm)	V_{OC} (mV)	J_{SC} (mA/cm^2)	FF (%)	Efficiency (%)
1	Gradient	613.7	32.56	81.93	16.37
2	Uniform	613.7	32.37	81.9	16.27

Table 4 Gradient doping and uniform doping of BSF layer (from 2e20 to 1e19) with the emitter layer thickness of 3 nm and textured TCO

S. No.	Emitter thickness (nm)	V_{OC} (mV)	J_{SC} (mA/cm^2)	FF (%)	Efficiency (%)
1	Gradient	613.7	32.56	81.93	16.37
2	Uniform	613.7	32.56	81.93	16.37

Table 5 Influence of front TCO layer texture emitter thickness of 3 nm and gradient doping

S. No.	Emitter thickness (nm)	V_{OC} (mV)	J_{SC} (mA/cm^2)	FF (%)	Efficiency (%)
1	Textured (pyramids)	613.7	32.56	81.93	16.37
2	Plane	612	29.77	81.7	14.89

the textured TCO layer minimizes the reflection losses as the reflected light again falls on the next texture of the surface and gets absorbed after multiple reflections into the solar cell. Table 5 shows the variation in output parameters due to the textured TCO Layer.

4 Discussion

- From the results of simulation 1, it is observed that by increasing the thickness of the emitter layer, the efficiency of the solar cell decreases. This is because of the increase in the recombination rate of electron–hole pairs.
- From the results of simulation 2, it is observed that with gradient doping of the emitter layer, the efficiency of the solar cell increases. This is because the

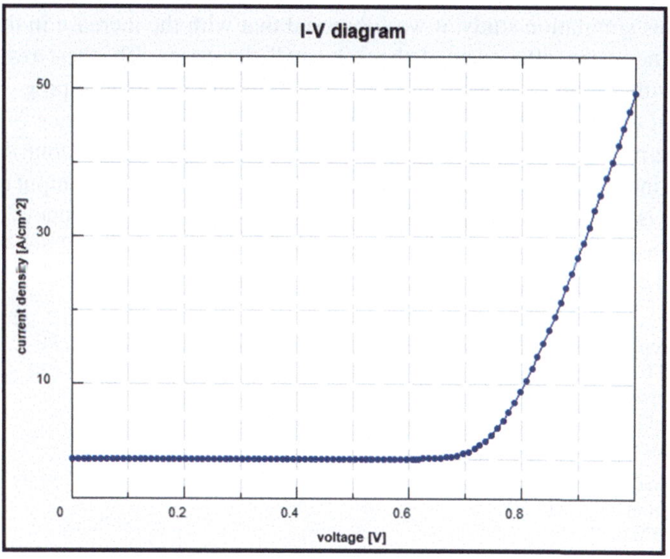

Fig. 3 *V–I* characteristics of the simulated bifacial solar cell

additional electric field coming from the gradient doping supports charge carrier separation.

- From the results of simulation 3, it is observed that with gradient doping of the back surface field layer, there is negligible to no effect has been seen on the efficiency of the solar cell.
- From the results of simulation 4, it is observed that with the textured transparent conductive oxide (TCO) layer, the efficiency of the solar cell increases because of the increased surface area and optical path which results in a decrease in the optical losses.
- The graph in Fig. 3 represents the *I–V* characteristics of the simulation model with the best results.

5 Conclusion

Fabrication of solar cells is a very expensive and resource-consuming process due to which we cannot use hit and trial methods to obtain the best output results. This simulation work can provide the optimum values of various parameters like layer thickness, doping levels, etc., in the solar cell for the best possible output results in terms of open-circuit voltage (V_{oc}), short circuit current (J_{sc}), fill factor (%FF), and efficiency (%η). Therefore, the cost of redundant fabrication of solar cells can be saved.

From this simulation study, it was observed that with the increase in the emitter layer thickness, the efficiency of the solar cell decreases. The best results were obtained with an emitter layer thickness of 3 nm with gradient doping. However, the gradient doping of the BSF layer has not affected the performance of the solar cell. Furthermore, the texturization of the front TCO layer has also contributed to a significant increase in solar cell performance. The best outcome of output efficiency of 16.37% is obtained with the fill factor of 81.93%. At this efficiency, the open-circuit voltage of 613.7 mV and short circuit current density of 32.56 A/cm^2 are observed. The best model parameters are given in the table below:

Doping of emitter	Doping of back surface field layer	TCO layer type	Emitter thickness (Nm)	V_{OC} (mV)	J_{SC} (mA/cm^2)	FF (%)	Efficiency (%)
Gradient (from 2e20 to 1e19)	Gradient (from 2e20 to 1e19)	Textured (pyramids < 111 >)	3	613.7	32.56	81.93	16.37

References

1. K. Vidyanandan, An overview of factors affecting the performance of solar PV systems. Energy Scan **27**(Feb), 2–8 (2017)
2. "Year End Review—Solar Power Target Reset to One Lakh MW;" https://pib.gov.in/newsite/PrintRelease.aspx?relid=133220. Accessed 16 May 2022
3. "Jawaharlal Nehru National Solar Mission (Phase I, II and III)—Policies—IEA." https://www.iea.org/policies/4916-jawaharlal-nehru-national-solar-mission-phase-i-ii-and-iii. Accessed 16 May 2022
4. "Current Status | Ministry of New and Renewable Energy, Government of India." https://mnre.gov.in/solar/current-status/. Accessed 16 May 2022
5. X. Wen, X. Zeng, W. Liao, Q. Lei, S. Yin, An approach for improving the carriers transport properties of a-Si:H/c-Si heterojunction solar cells with efficiency of more than 27%. Sol. Energy **96**, 168–176 (2013). https://doi.org/10.1016/J.SOLENER.2013.07.019
6. Y. Yao, X. Xu, X. Zhang, H. Zhou, X. Gu, S. Xiao, Enhanced efficiency in bifacial HIT solar cells by gradient doping with AFORS-HET simulation. Mater. Sci. Semicond. Process. **77**, 16–23 (2018). https://doi.org/10.1016/J.MSSP.2018.01.009
7. R. Varache, C. Leendertz, ME. Gueunier-Farret, J. Haschke, D. Muñoz, L. Korte, Investigation of selective junctions using a newly developed tunnel current model for solar cell applications. Solar. Energy. Mater. Solar. Cells. 14114–14123 (2015). https://doi.org/10.1016/j.solmat.2015.05.014

An Investigation on the Effect of Charging Current on Capacity, Coulombic Efficiency, and Energy Density of Commercial Lithium-Ion Polymer Cells

Vijaya and Pankaj Kalita

Abstract The present study provides insights on the charging of batteries at varying current levels aimed at implementation in high altitude areas that require specialized charging protocols especially for specific military applications. To examine the effect of levels of charging current on battery performance, commercial Lithium-ion Polymer (LiPo) cells are subjected to Constant Current Constant Voltage (CCCV) charging at varying current levels for 500 cycles. The analysis of results indicates the significance of the sequence of charging current on cycle life. In the LiPo cell subjected to relatively higher levels of charging current in the beginning, the onset of degradation occurs 50 cycles in advance. The extent of deterioration also determines the charging time. After the onset of degradation, capacity fade is rapid and charging is slower regardless of the charging current. The onset of capacity degradation also corresponds to the onset of degradation of energy density (ED) and Coulombic efficiency (CE). At the end of 500 cycles, the two LiPo cells retained 81% and 90% of initial capacity, respectively, which suggests that with the right selection of the sequence of charging currents, this novel method of charging strategy can be successfully implemented without shortening the battery cycle life.

Keywords Energy storage · Li-ion battery · Charging protocols · SOH · CCCV charging · Capacity degradation

1 Introduction

At present, rechargeable Li-ion batteries find diverse applications in various spheres of living, ranging from but not limited to renewable energy storage, portable electronics, wearables, agriculture power tools, e-mobility, and space missions. The market share of batteries is expected to expand further in the coming years. In this regard, proper understanding of the impact of different charging strategies on battery performance is vital for effective and safe operation of Li-ion batteries.

Vijaya · P. Kalita (✉)
School of Energy Science and Engineering, Indian Institute of Technology Guwahati, Guwahati, Assam 781039, India
e-mail: pankajk@iitg.ac.in

V. S. Moholkar et al. (eds.), *Sustainable Energy Generation and Storage*,
https://doi.org/10.1007/978-981-99-2088-4_9

1.1 Principle Components and Basic Working Principle of Li-Ion Battery

Commercial Li-ion batteries mainly consist of graphite anode, lithium metal oxide cathode, typically lithium cobalt oxide ($LiCoO_2$referred to as LCO), lithium nickel manganese cobalt oxide (NMC), lithium iron phosphate ($LiFePO_4$), and non-aqueous electrolytes, predominantly Lithum hexafluorophosphate ($LiPF_6$) dissolved in carbonates such as ethylene carbonate, propylene carbonate, diethyl carbonate, and diethyl carbonate [1, 2]. Besides non-aqueous electrolytes, polymer electrolytes such as poly (ethylene oxide), poly (acrylonitrile), poly (vinylidene fluoride), and poly (methyl methacrylate) have been increasingly finding applications in Li-ion batteries owing to good ionic conductivity, chemical stability, mechanical strength, easy processability to form thin films with good viscoelasticity, low flammability, and hence better safety [3, 4]. Polymer electrolytes were first proposed to be used in rechargeable batteries in 1978 by Armanda et al. [5] and have been widely used in batteries ever since. The main difference between a Li-ion battery and a LiPo battery is in the type of electrolyte used. While the former uses a liquid electrolyte, the latter uses polymer electrolyte such as poly (ethylene oxide), PEO, and a lithium salt [6].

The working of the Li-ion battery is governed by the principles of redox reaction and is dependent on the back-and-forth movement of Li ions between the electrodes through the electrolyte, referred to as rocking chair movement [2, 7, 8]. In a conventional Li-ion battery with LCO cathode and graphite anode, during charging, Li ions split from LCO cathode and are intercalated into the layers of graphite anode. The release of Li ions is accompanied by simultaneous release of electrons as a result of oxidation of Co (III) in cathode to Co (IV). Discharging process involves the movement of Li ions back to host framework from the anode [9].

The half-reactions and overall reaction are given below [10]. The forward direction represents reactions during discharging and the reverse represents charging.

$$\text{Anode half reaction}: \quad \mathbf{LiC_6 \leftrightarrows 6C} + \boldsymbol{x}\mathbf{Li^+} + \mathbf{xe^-} \tag{1}$$

$$\text{Cathode half reaction}: \quad \boldsymbol{x}\mathbf{Li^+} + \boldsymbol{x}\mathbf{e^-} + \mathbf{Li_{1-x}CoO_2 \leftrightarrows LiCoO_2} \tag{2}$$

$$\text{Overall reaction}: \quad \mathbf{LiC_6 + CoO_2 \leftrightarrows C_6} + \text{Li } \mathbf{CoO_2} \tag{3}$$

Figure 1 illustrates the working of a battery during discharging. Ion transport during charge–discharge process mainly comprises 3 key mechanisms, viz., (1) Movement through the electrodes, (2) Movement through the electrode–electrolyte interface, and (3) Movement through the electrolyte among which ionic diffusion through the electrodes is a major rate-limiting step that determines the battery power performance [11].

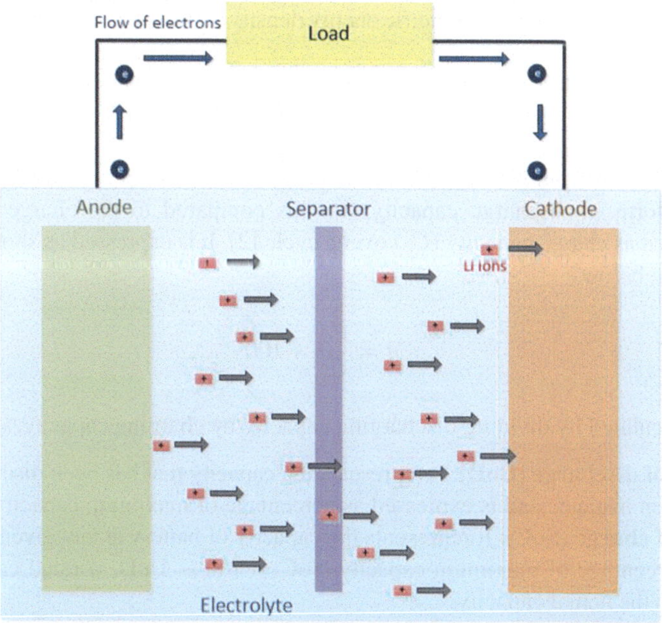

Fig. 1 Schematic representation of working of a Li-ion battery

1.2 Terminologies and Performance Metrics

Choice of battery for various applications depends majorly on battery performance metrics such as capacity, energy density, power density, and cycle life and factors such as cost and safety across operating range. Some of the commonly used performance metrics and terminologies related to battery are explained below.

- **Capacity**: Capacity of a battery is the total charge available at the battery and is represented as ampere- hours (Ah) [12]. It is an indicator of time for which a battery can power a load before running flat when discharged at specified rate. For example, a battery with a capacity of 2 Ah will theoretically last for 1 h when the discharge current is 2 A.
- **C-rate**: It is the measure of rate at which battery is discharged relative to its full capacity. For a battery with capacity of 2 Ah, a discharge at 1C rate would mean discharge current of 2 A discharges the battery completely in 1 h.
- **Energy density (ED)**: Mathematically, ED is the product of capacity and voltage and is expressed in Wh/Kg or Wh/L depending on whether it is gravimetric energy density or volumetric energy density, respectively. It is an indication of the amount of energy that can be stored per unit mass or volume. The former is called gravimetric energy density, and the latter is called volumetric energy density. The weight of battery to meet particular capacity is determined by its gravimetric

energy density whereas volumetric energy density dictates the space requirements of a battery.

- **Power density**: It is the maximum deliverable power of a battery. It is dependent on kinetics of charge migration and transfer of the device [2]. The SI unit of power density is W/kg.
- **Coulombic efficiency (CE)**: It is a measure of recoverable charge often measured in the form of discharge capacity (C_d) as compared to the charge injected, measured as charge capacity (C_c) over a cycle [2]. It is expressed as shown in the equation below.

$$\eta = \frac{C_d}{C_c} \times 100$$

It is calculated by dividing discharging capacity by charging capacity.

- **Depth of discharge (DoD)**: It represents the capacity that has been discharged at any given instance and is expressed as percentage of maximum capacity [12].
- **State of charge (SoC)**: It represents the capacity of battery at any given instance as a percentage of maximum capacity. SoC $=100-$ DoD, if rated capacity is equal to the actual capacity.
- **Cycle life:** It represents the total number of charge–discharge cycles over which a battery can effectively operate. According to the US Advanced Battery Consortium, a battery is said to be at the end of its life when the specific capacity drops to 80% of its initial capacity [13–15].

1.3 Charging Standards and Charging Protocols

Several DC and AC charging standards are followed worldwide. SAE and IEEE (USA), IEC (Europe), CHAdeMO (Japan), and GuoBiao (China) are the most widely used among all [16, 17]. The charging standards determine the maximum output voltage, current level and power level, communication interface between the vehicle and the charging device/grid, types of sockets, connectors, plugs, and their location (on-board, off-board).

The various charging schemes can be broadly classified as passive optimal charging and active optimal charging protocol. Passive charging protocols run on simple algorithms and are easy to implement in battery management systems (BMS). Unlike passive charging, active charging protocols are model based and couple charging time with vital parameters of battery health to optimize charging [18]. Active charging protocols can be further classified into empirical and physics-based model schemes.

Some of the conventionally used passive charging protocols are described below.

Constant Current (CC) charging: In the CC mode of charging, the battery is charged with constant current while the voltage keeps rising. One of the major challenges in this scheme of charging is to achieve 100% state of charge while maintaining the voltage within safety limits.

Constant Voltage (CV) charging: In this mode of charging, the battery is charged with a constant voltage while the current keeps rising. High starting current and temperature rise during charging are the major drawbacks of this mode.

Constant Power (CP) charging: In this mode of charging, varying values of current are applied so as to keep the power constant.

Constant Current Constant Voltage (CCCV) Charging: In the CCCV mode of charging, the battery is charged with a constant current until a preset cutoff voltage is reached after which it is charged at the constant voltage until a current cutoff is reached [19].

CCCV charging is further optimized to boost charging, pulse charging and multi-stage constant current charging (MCC), varying current decay charging (VCD), and constant power constant voltage charging (CPCV) [20].

The major bottleneck to fast charging is capacity degradation at higher C-rates. The charging time can be considerably reduced by charging the battery at higher currents but this causes temperature rise, affects the overall life of the battery, and in some cases, may cause permanent damage [21].

There are mainly four approaches to understanding battery degradation, viz. (i) factors that contribute to degradation, (ii) degradation mechanisms, (iii) degradation modes, and (iv) observable effects of degradation as given in [22, 23] (Table 1).

Temperature has a significant effect on the rate of chemical reactions in the battery and directly affects the battery aging [35, 36]. Rise in temperature beyond a certain limit results in thermal runaway reaction which poses the risk of fire hazards and is a serious safety concern [11]. In recent years, there have been several simulation-based, electrochemical, and thermal model-based studies [37, 38] to optimize conventional charging strategies and to develop new strategies with an aim to reduce charging time without hampering battery safety and cycle life. In addition, several experimental

Table 1 Approaches to analyzing battery degradation

Factors causing degradation	Degradation mechanisms		Degradation modes	Observable effects of degradation
	Primary	Secondary		
Temperature [24–31] SoC [20, 32–34] Load profile	SEI layer formation Particle fracture Li plating and dendrite growth Structural disordering	Graphite exfoliation Pore blockage Island formation Electrolyte decomposition Transition metal dissolution Gas formation	Loss of active material Loss of lithium inventory Stoichiometric drift Impedance change	Capacity fade Power fade

studies on various Li-ion chemistries performed over 1000 cycles provide insights on the influence of multiple factors such as battery chemistry, charging protocol, and temperature on the life of battery [23–26].

Ansean et al. [39] demonstrated a multistage CCCV technique on lithium iron phosphate (LFP) batteries in which battery was subjected to CC charging at 4C until cutoff voltage of 3.6 V was reached. Thereafter, the battery was charged at 1C up to 3.6 V and the constant voltage of 3.6 V is maintained for 5 min. The proposed protocol resulted in charging of battery from 0 to 100% SOC within 20 min even after 4500 cycles with capacity degradation of 83%. Keil et al. [40] demonstrated the influence of different charging protocols, viz. CCCV charging, pulse charging and boost charging on the cycle life, charging time, capacity utilization, and energy efficiency of three types of high-power 18,650 Li-ion cells. The results indicate higher influence of charging current on cycle life as compared to discharging current. The authors suggest reducing charging currents at high SoC to prevent lithium plating and extend the cycle life. Soto et al. [41] undertook capacity, open circuit voltage, and internal resistance characterization of three different 18,650 Li-ion cells. The discharge capacity of cells subjected to CCCV charging and discharging showed a decreasing trend with increasing C-rates at the end of CC stage while the cells showed higher discharge capacity at higher C-rate at the end of CV stage. The higher value of capacity at higher C-rates is ascribed to rise in temperature, and hence, the authors recommend discharge to be carried out at lower C-rates (< C/3). Wang et al. [42] demonstrated the influence of temperature and charging protocol on Li-ion batteries subjected to CCCV and CC charging at 25 and 60 °C for 1000 cycles. Batteries subjected to CC charging at 25 °C degraded faster as compared to batteries subjected to other treatments.

In general, the impact of charging protocol was found to vary considerably with battery chemistry. Hence, it is important to study and document the influence of various charging protocols on the performance of different battery chemistries. In this paper, the effect of varying levels of charging current on the cycle life of battery is discussed by analyzing the performance parameters such as capacity, energy density, and Coulombic efficiency along with SOC, SOH, and charging time. The novel method of charging addresses the effect of charging current on battery cycle life considering a scenario where users have the provision to charge batteries at various current levels; typically, a scenario where different current levels are marked as notches on a knob-based charger and the user can select the level of charging by rotating the knob. The proposed scheme of charging will also be useful for application of batteries in high altitude areas where the temperature drops below the optimal operating range of batteries (15–35 °C) during the winter, posing a challenge to charging. The degradation of battery performance is rapid with decreasing ambient temperature which calls for a specialized charging scheme [36, 43–45].

2 Experimental Method

To understand the influence of charging protocol on cell performance, two prismatic LiPo cells (shown in Fig. 2) were subjected to CCCV charging at 25 ± 3 °C for about 6 months. The cells used in this study are commonly used in drones and other robotic applications. Specifications of the cells are given in Table 2. Charging and discharging were carried out using cell charge and discharge system (Make: Chroma, Model: 17,011).

Fig. 2 The LiPo cells that were tested in this study

Table 2 Battery specifications

Battery	Li polymer battery
Manufacturer	SkyCell
Chemistry	Lithium cobalt oxide
Model	RKI-6000
Dimensions	115 mm × 35 mm × 7 mm
Nominal capacity	2200 mAh
Nominal voltage	3.7 V
Maximum charging voltage	4.2 V
Maximum charging current	1 A
Maximum discharge current	–
Cycle life	Full capacity for up to 1000 charge cycles

Table 3 CCCV charging protocol

	Setting		Cutoff condition
Protocol	Voltage (V)	Current (A)	
CCCV charging	4.2	**I**	I < 0.06 A
Rest			Time ≥ 1200 s
CC discharging		1.1	V < 2.8 V
Rest			Time ≥ 1200 s

Table 4 Charging current for CCCV charging

Number of cycles	50	100	150	200	250	300	350	400	450	500
LiPo 1	2.2 A	1.6 A	0.4 A	1.2 A	0.8 A	2 A	0.6 A	1 A	1.8 A	1.4 A
LiPo 2	2.2 A	1.6 A	1.2 A	0.4 A	2 A	0.8 A	1 A	1.4 A	0.6 A	1.8 A

In order to limit the effect of discharge current on cycle life, the cells were discharged at 0.5C, i.e., 1.1 A, under Constant Current (CC) discharging mode. The cutoff voltage was set to 2.8 V for CC discharge.

The cells were subjected to continuous charging (CCCV) and discharging (CC) with 20 min of rest between both the processes for 500 cycles. The details of the protocols are given in Table 3. Various values of **I** mentioned in CCCV charging are mentioned in Table 4. The two LiPo batteries are referred to as Lithium Polymer 1 (LiPo 1) and Lithium Polymer 2 (LiPo 2). To understand the influence of charging current on battery performance and to ensure the observed results are not due to chance alone, the values of **I** were assigned randomly for successive cycles [40, 41].

3 Results

Figure 3 shows the discharge capacity of LiPo 1 and LiPo 2 at different charging currents. Figure 3c represents the overall discharge capacity profile of LiPo 1 and LiPo 2 in the order of charging currents mentioned in Table 4, i.e., first 50 cycles correspond to discharge capacity when battery is charged at 2.2 A, next 50 cycles correspond to discharge capacity when charging current is 1.6 A, etc. Although charging currents are different for LiPo 1 and LiPo 2, the overall discharge capacity profile indicates a gradual decline in capacity after multiple charge–discharge cycles. From Fig. 3c, it is observed that the discharge capacity of LiPo 1 and LiPo 2 remain around rated 2.2 Ah for 400 cycles and 350 cycles, respectively. Thereafter, it declines rapidly below rated capacity. The corresponding cycle number (350 and 400 onwards in case of LiPo 1 and LiPo 2, respectively) is referred to as the point of onset of capacity degradation.

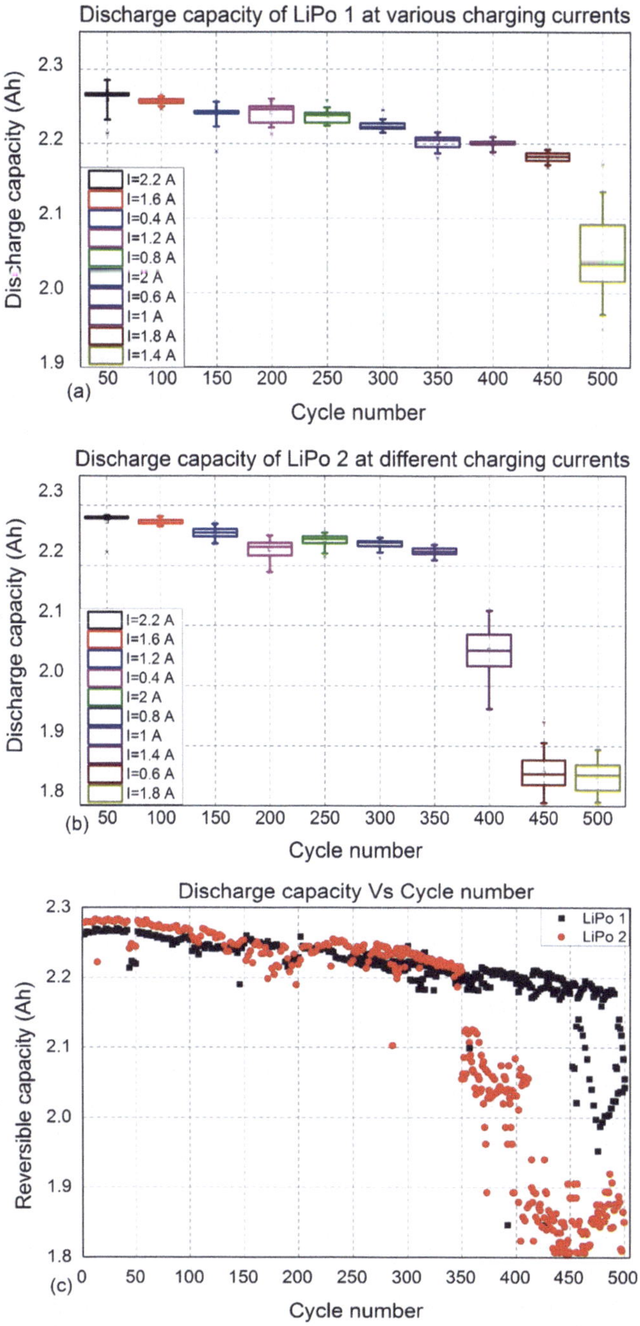

Fig. 3 **a, b** Discharge capacity of LiPo 1 and LiPo 2 at different charging currents, respectively. **c** Overall discharge capacity profile of the cells at various charging currents for 500 cycles

Table 5 Rate of capacity degradation of batteries over 500 cycles

Initial cycle number	Final cycle number	Initial capacity (Ah)		Final capacity (Ah)		Rate of capacity degradation (%)	
		B1	B2	B1	B2	B1	B2
2	500	2.26	2.28	2.04	1.85	– 9.66	– 18.76

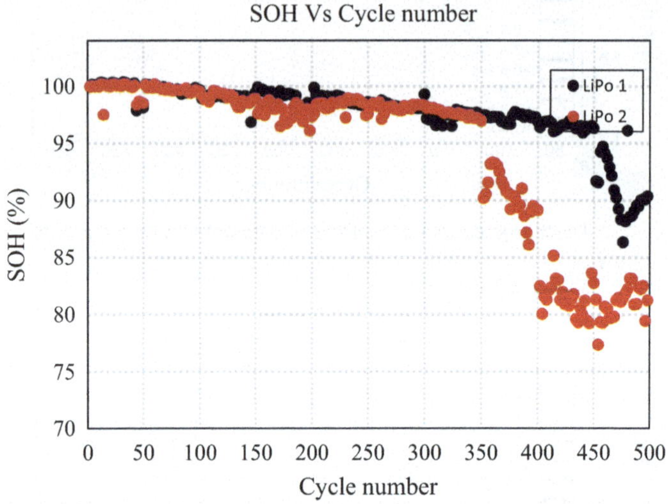

Fig. 4 Plot of SOH versus cycle number of LiPo 1 and LiPo 2

From the rate of degradation of capacity in Table 5 and the plot of State Of Health (SOH) versus cycle life in Fig. 4, it is observed that the overall rate of capacity degradation is higher in LiPo 2 (– 18.76%) as compared to LiPo 1 (– 9.66%).

LiPo1 and LiPo2 are subjected to identical levels of charging current for first 100 cycles. Thereafter, the sum total of charging current is same but the sequence is different. Between 100 and 250 cycles, LiPo1 was subjected to lower levels of charging current (0.4 A, 1.2 A, and 0.8 A) overall as compared to LiPo 2 (1.2 A, 0.4 A, 2 A) leading to the advancement of degradation beginning from 300th cycles onwards. From these results, it is inferred that the sequence of charging current plays an important role in determining the overall cycle life of the batteries. Further, it is also observed that the capacity fade is irreversible in nature, i.e., charging at lower currents does not slow down the rate of capacity fade after the onset of degradation.

Further, from the plot of SOC versus Charging time in Fig. 5, it is observed that the time required to fully charge the battery drastically varies after the onset of degradation and charging at higher currents does not ensure faster charging after the battery capacity deteriorates. For instance, the charging time of LiPo 1 at 1.2 A (02 h 40 min and 07 s) is lesser than at 1.4 A (03 h 07 min and 57 s). Similarly, the charging time of LiPo 2 at 1.8 A > at 1.6 A, 1.4 A, 1.2 A, 1 A, 0.8 A, and 0.6 A.

The higher the extent of degradation, the longer the charging time as observed in the case of charging at 1.8 A; LiPo 2 requires double the time (04 h, 07 min, and 31 s) to be fully charged as compared to LiPo 1 (02 h, 42 min, and 18 s).

CE of the batteries remains above 99% for most parts of 500 cycles as seen from Fig. 6. CE of LiPo 2 is observed to be more scattered than LiPo 1, and the scattering is prominent beyond 350 cycles, i.e., after the onset of capacity degradation. The uneven scatterings of CE observed in LiPo 1 and LiPo 2 are indications of irregularities in ion intercalation–deintercalation at higher cycles.

Volumetric energy density (ED) is reported considering its significance in applications such as portable electronics and electric vehicles where storage space is a major constraint [46]. ED is determined by dividing the energy stored per charge–discharge cycle which is obtained from chroma charge–discharge tester by the cell

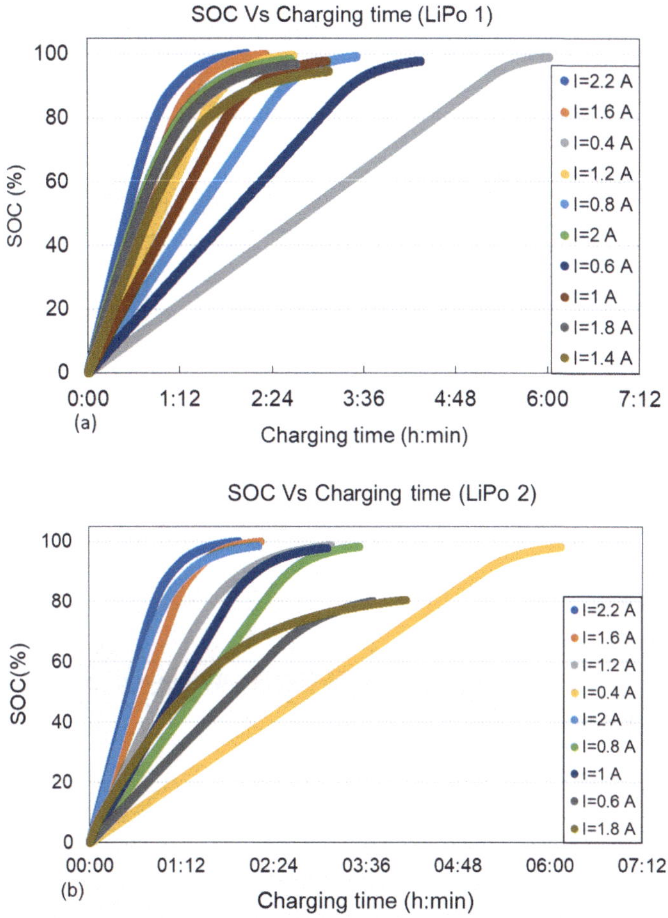

Fig. 5 **a**, **b** Plot of SOC versus charging time of LiPo 1 and LiPo 2

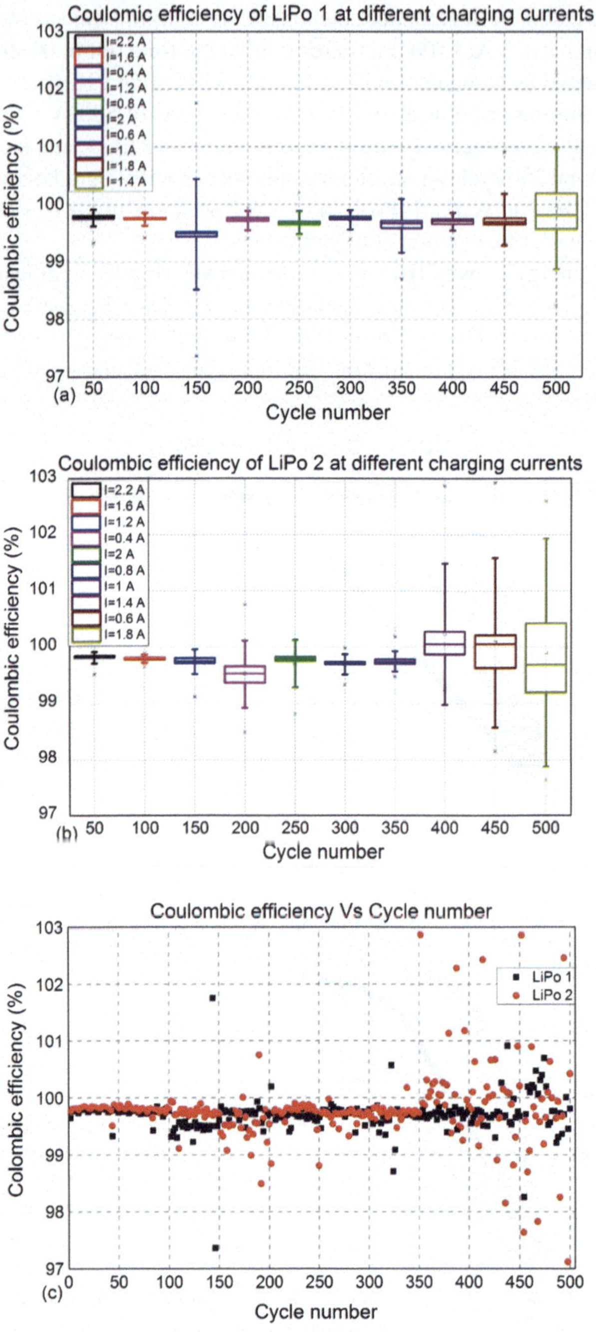

Fig. 6 **a, b** Coulombic efficiency of LiPo 1 and LiPo 2 at different charging currents, respectively. **c** Overall Coulombic efficiency profile of the cells at various charging currents for 500 cycles

dimensions (115 mm × 35 mm × 7 mm). From Fig. 7, it is observed that ED of cells is directly proportional to charging currents until the onset of capacity degradation. Thereafter, higher charging currents do not translate to higher ED.

4 Conclusion

The effect of charging the batteries at different current levels over its cycle life was investigated by a long-term experimental study. The significance of the sequence of charging current is analyzed in this paper by considering the battery capacity, Coulombic efficiency, energy density, SOC, SOH, and charging time. The impact of levels of charging current on battery performance is not evident up to the onset of degradation which occurs after 350 and 400 cycles, respectively, for LiPo 1 and LiPo 2. However, after the onset, the rate of degradation is rapid and charging is slower irrespective of the charging current. The onset of capacity degradation is also reflected in deterioration in SOH, Coulombic efficiency, and energy density of the battery. Past studies attribute capacity degradation to active material loss, internal resistance, Li plating at the surface of electrodes, and other primary and secondary reactions listed in [47–50]. The influence of each of these factors needs to be studied further.

By investigating the performance of LiPo cells charged at different current levels for 50 cycles each for a total of 500 cycles, this study provides insights into charging batteries at varying current levels which will be useful to recommend a best practice for battery charging in high altitude regions which requires specialized charging strategy. By choosing the right sequence of charging currents, this novel method of charging can be implemented in applications that require charging at higher currents during certain occasions (e.g., batteries in military surveillance drones in high altitude, cold temperature regions) without compromising on the cycle life of the battery.

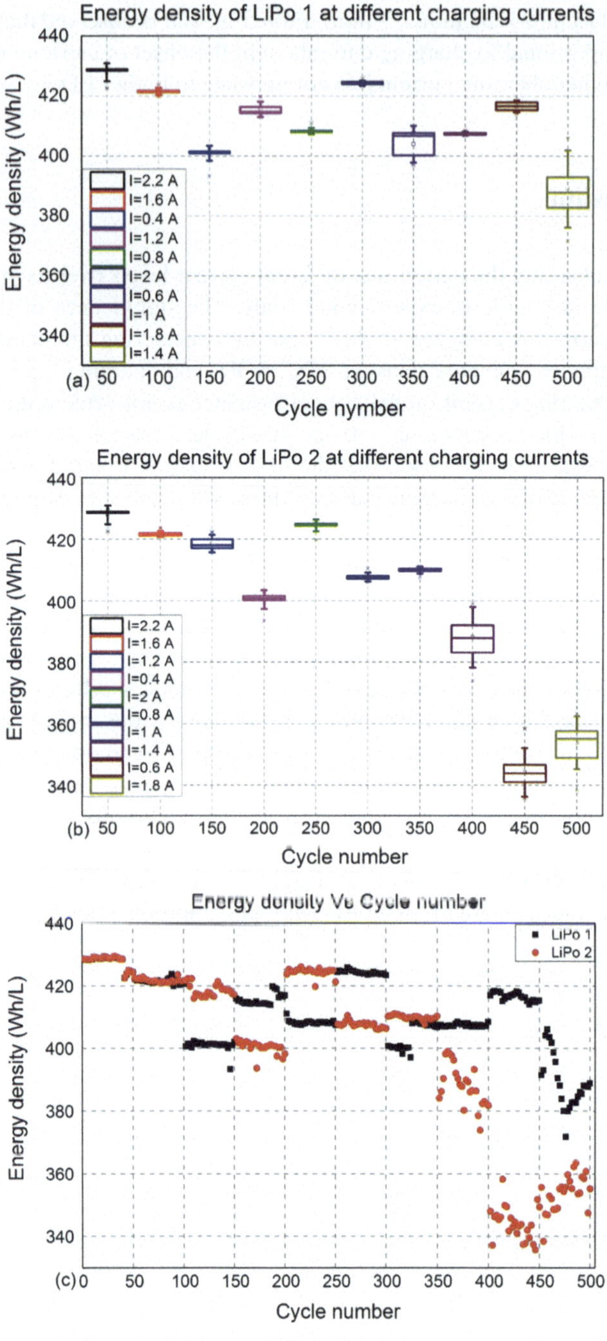

Fig. 7 **a**, **b** energy density of LiPo 1 and LiPo 2 at different charging currents, respectively. **c** Overall energy density profile of the cells at various charging currents for 500 cycles

Acknowledgements This work was supported by Indian Institute of Technology Guwahati under the startup top-up Grant [Grant No. R&D/SG ITOP UP12018-19/08].

Author declaration: The authors declare no conflicts of interest.

Data Availability The data that support the findings of this study are available from the corresponding author upon reasonable request.

References

1. Q. Li, J. Chen, L. Fan, X. Kong, Y. Lu, Progress in electrolytes for rechargeable Li-based batteries and beyond. Green Energy Environ. **1**(1), 18–42 (2016). https://doi.org/10.1016/j.gee.2016.04.006
2. R. Borah, F.R. Hughson, J. Johnston, T. Nann, On battery materials and methods. Mater. Today Adv. **6**, 100046 (2020). https://doi.org/10.1016/j.mtadv.2019.100046
3. A. Arya, A.L. Sharma, Polymer electrolytes for lithium ion batteries: a critical study. Ionics **23**(3) (2017). https://doi.org/10.1007/s11581-016-1908-6
4. V. Deshpande et al., Journal of physics: energy 2020 roadmap on solid-state batteries. J. Phys. Energy **2**(3), 32008 (2020)
5. N. Boaretto, L. Meabe, M. Martinez-Ibañez, M. Armand, H. Zhang, Review—polymer electrolytes for rechargeable batteries: from nanocomposite to nanohybrid. J. Electrochem. Soc. **167**(7), 070524 (2020). https://doi.org/10.1149/1945-7111/ab7221
6. B. Scrosati, F. Croce, S. Panero, Progress in lithium polymer battery R&D. J. Power Sources **100**(1–2), 93–100 (2001). https://doi.org/10.1016/S0378-7753(01)00886-2
7. S. Sarkar, Y.C. Jou, Adaptive control of the reverse link in cdma2000. Int. J. Wireless Inf. Networks **9**(1), 55–70 (2002). https://doi.org/10.1023/A:1014832402210
8. J.M. Tarascon, D. Guyomard, The Li_1+xMn_2O_4/C rocking-chair system: a review. Electrochim. Acta **38**(9), 1221–1231 (1993). https://doi.org/10.1016/0013-4686(93)80053-3
9. J. Xie, Y.C. Lu, A retrospective on lithium-ion batteries. Nat. Commun. **11**(1), 9–12 (2020). https://doi.org/10.1038/s41467-020-16259-9
10. J. Xu, H.R. Thomas, R.W. Francis, K.R. Lum, J. Wang, B. Liang, A review of processes and technologies for the recycling of lithium-ion secondary batteries **177**, 512–527 (2008). https://doi.org/10.1016/j.jpowsour.2007.11.074
11. N. Nitta, F. Wu, J.T. Lee, G. Yushin, Li-ion battery materials: present and future. Mater. Today **18**(5), 252–264 (2015). https://doi.org/10.1016/j.mattod.2014.10.040
12. M.J. Piernas Muñoz, E. Castillo Martínez, Introduction to batteries. SpringerBriefs Appl. Sci. Technol. (December), 1–8 (2018). https://doi.org/10.1007/978-3-319-91488-6_1
13. Y. Zhao et al., A review on battery market trends, second-life reuse, and recycling. Sustain. Chem. **2**(1), 167–205 (2021). https://doi.org/10.3390/suschem2010011
14. G. Hunt, Electric vehicle battery test procedures revision 2. Idaho Nat. Eng. Lab. (INEL) (January), 1–40 (1996)
15. Idaho National Laboratory. Battery Calendar Life Estimator Manual no. October (2012)
16. H.S. Das, M.M. Rahman, S. Li, C.W. Tan, Electric vehicles standards, charging infrastructure, and impact on grid integration: a technological review. Renew. Sustain. Energy Rev. **120**(February 2019) (2020). https://doi.org/10.1016/j.rser.2019.109618
17. CHAdeMO, *IEEE Standard Technical Specifications of a DC Quick Charger for Use with Electric Vehicles IEEE Vehicular Technology Society IEEE Standard Technical Specifications of a DC Quick Charger for Use with Electric Vehicles* (2015)
18. Y. Gao, X. Zhang, Q. Cheng, B. Guo, J. Yang, Classification and review of the charging strategies for commercial lithium-ion batteries. IEEE Access **7**, 43511–43524 (2019). https://doi.org/10.1109/ACCESS.2019.2906117

19. B. Tarm A. Fayed, An overview of the fundamentals of battery chargers. Midwest Symp. Circ. Syst. (October), 16–19 (2016). https://doi.org/10.1109/MWSCAS.2016.7870048
20. M. Adaikkappan, N. Sathiyamoorthy, Modeling, state of charge estimation, and charging of lithium-ion battery in electric vehicle: a review. Int. J. Energy Res. **46**(3), 2141–2165 (2022). https://doi.org/10.1002/er.7339
21. C. Suarez, W. Martinez, Fast and ultra-fast charging for battery electric vehicles—a review. 2019 IEEE Energy Convers. Congr. Exposition ECCE **2019**, 569–575 (2019). https://doi.org/10.1109/ECCE.2019.8912594
22. J.S. Edge et al., Lithium ion battery degradation: what you need to know. Phys. Chem. Chem. Phys. **23**(14), 8200–8221 (2021). https://doi.org/10.1039/d1cp00359c
23. M. Woody, M. Arbabzadeh, G.M. Lewis, G.A. Keoleian, A. Stefanopoulou, Strategies to limit degradation and maximize Li-ion battery service lifetime—critical review and guidance for stakeholders. J. Energy Storage. **28** (2020). https://doi.org/10.1016/j.est.2020.101231
24. Y. Xiong, Strategy and temperature affect of the charging of lithium-ion battery management system, in *Proceedings of the 2015 International conference on Applied Science and Engineering Innovation*, vol. 12, no. Asei (2015), pp. 1796–1799. https://doi.org/10.2991/asei-15.2015.357
25. S. Ma et al., Temperature effect and thermal impact in lithium-ion batteries: a review. Prog. Nat. Sci. Mater. Int. **28**(6), 653–666 (2018). https://doi.org/10.1016/j.pnsc.2018.11.002
26. A. Tung, M. Gan, Study the thermal effect on low cost lithium ions battery. Int. J. Appl. Eng. Res. **13**(1), 9154–9158 (2018)
27. F. Leng, C.M. Tan, M. Pecht, Effect of temperature on the aging rate of li ion battery operating above room temperature. Sci. Rep. **5**, 1–12 (2015). https://doi.org/10.1038/srep12967
28. J.B. Robinson et al., Non-uniform temperature distribution in Li-ion batteries during discharge—a combined thermal imaging, X-ray micro-tomography and electrochemical impedance approach. J. Power Sources **252**, 51–57 (2014). https://doi.org/10.1016/j.jpowsour.2013.11.059
29. H.J. Kim, J.H. Lee, D.H. Baek, J.K. Lee, A study on thermal performance of batteries using thermal imaging and infrared radiation. J. Ind. Eng. Chem. **45**, 360–365 (2017). https://doi.org/10.1016/j.jiec.2016.10.003
30. P. Wang et al., Real-time monitoring of internal temperature evolution of the lithium-ion coin cell battery during the charge and discharge process. Extreme Mech Lett **9**, 459–466 (2016). https://doi.org/10.1016/j.eml.2016.03.013
31. S. Chacko, Y.M. Chung, Thermal modelling of Li-ion polymer battery for electric vehicle drive cycles. J. Power Sources **213**, 296–303 (2012). https://doi.org/10.1016/j.jpowsour.2012.04.015
32. T.M.N. Bui, M. Sheikh, T.Q. Dinh, A. Gupta, D.W. Widanalage, J. Marco, A study of reduced battery degradation through state-of-charge pre-conditioning for vehicle-to-grid operations. IEEE Access **9**, 155871–155896 (2021). https://doi.org/10.1109/ACCESS.2021.3128774
33. Y. Zhang, R. Xiong, H. He, X. Qu, M. Pecht, State of charge-dependent aging mechanisms in graphite/Li(NiCoAl)O_2 cells: capacity loss modeling and remaining useful life prediction. Appl. Energy. **255** (2019). https://doi.org/10.1016/j.apenergy.2019.113818
34. E. Wikner, T. Thiringer, Extending battery lifetime by avoiding high SOC. Appl. Scie. (Switzerland) **8**(10) (2018). https://doi.org/10.3390/app8101825
35. L. Batteries, *Energies Analysis of the Effect of the Variable Charging* (2019)
36. A. Tomaszewska et al., Lithium-ion battery fast charging: a review. eTransportation **1**, 100011 (2019). https://doi.org/10.1016/j.etran.2019.100011
37. Z. Guo, B.Y. Liaw, X. Qiu, L. Gao, C. Zhang, Optimal charging method for lithium ion batteries using a universal voltage protocol accommodating aging. J. Power Sources **274**, 957–964 (2015). https://doi.org/10.1016/j.jpowsour.2014.10.185
38. G. Chen, Z. Liu, H. Su, An optimal fast-charging strategy for lithium-ion batteries via an electrochemical–thermal model with intercalation-induced stresses and film growth. Energies (Basel) **13**(9) (2020). https://doi.org/10.3390/en13092388
39. D. Anseán, M. González, J.C. Viera, V.M. García, C. Blanco, M. Valledor, Fast charging technique for high power lithium iron phosphate batteries: a cycle life analysis. J. Power Sources **239**, 9–15 (2013). https://doi.org/10.1016/j.jpowsour.2013.03.044

40. P. Keil, A. Jossen, Charging protocols for lithium-ion batteries and their impact on cycle life-An experimental study with different 18650 high-power cells. J. Energy Storage **6**, 125–141 (2016). https://doi.org/10.1016/j.est.2016.02.005
41. A. Soto, A. Berrueta, P. Sanchis, A. Ursua, Analysis of the main battery characterization techniques and experimental comparison of commercial 18650 Li-ion cells. In *Proceedings—2019 IEEE International Conference on Environment and Electrical Engineering and 2019 IEEE Industrial and Commercial Power Systems Europe, EEEIC/I and CPS Europe 2019*, pp. 1–6, 2019. https://doi.org/10.1109/EEEIC.2019.8783862
42. H. Wang, S. Frisco, E. Gottlieb, R. Yuan, J.F. Whitacre, Capacity degradation in commercial Li-ion cells: the effects of charge protocol and temperature. J. Power Sources **426**(March), 67–73 (2019). https://doi.org/10.1016/j.jpowsour.2019.04.034
43. J.R.M. Delos Reyes, R.V. Parsons, R. Hoemsen, Winter happens: the effect of ambient temperature on the travel range of electric vehicles. IEEE Trans. Veh. Technol. **65**(6), 4016–4022 (2016). https://doi.org/10.1109/TVT.2016.2544178
44. T. Wang et al., Performance of plug-in hybrid electric vehicle under low temperature condition and economy analysis of battery pre-heating. J. Power Sources **401**(September), 245–254 (2018). https://doi.org/10.1016/j.jpowsour.2018.08.093
45. C. Vidal, S. Member, O. Gross, R. Gu, P. Kollmeyer, A. Emadi, xEV Li-Ion battery low-temperature effects—review. **68**(5), 4560–4572 (2019)
46. M. Ue, K. Sakaushi, K. Uosaki, Basic knowledge in battery research bridging the gap between academia and industry. Mater. Horiz. **7**(8), 1937–1954 (2020). https://doi.org/10.1039/d0mh00067a
47. S.S. Choi, H.S. Lim, Factors that affect cycle-life and possible degradation mechanisms of a Li-ion cell based on $LiCoO_2$. J. Power Sources **111**(1), 130–136 (2002). https://doi.org/10.1016/S0378-7753(02)00305-1
48. S. Nowak, M. Winter, Review—chemical analysis for a better understanding of aging and degradation mechanisms of non-aqueous electrolytes for lithium ion batteries: method development, application and lessons learned. J. Electrochem. Soc. **162**(14), A2500–A2508 (2015). https://doi.org/10.1149/2.0121514jes
49. R. Hausbrand et al., Fundamental degradation mechanisms of layered oxide Li-ion battery cathode materials: methodology, insights and novel approaches. Mater. Sci. Eng. B: Solid-State Mater. Adv. Technol. **192**(C), 3–25 (2015). https://doi.org/10.1016/j.mseb.2014.11.014
50. S.K. Jung et al., Understanding the degradation mechanisms of $LiNi_{0.5}Co_{0.2}Mn_{0.3}O_2$ cathode material in lithium ion batteries. Adv. Energy Mater. **4**(1), 1–7 (2014). https://doi.org/10.1002/aenm.201300787

Modelling of p-a-Si:H/i-a-Si:H/(n)c-Si Silicon Solar Cells by AFORS-HET Software

Juhi Kumari, Rahul, and Pratima Agarwal

Abstract We have performed simulation studies using AFORS-HET software to understand how dopant concentration (i.e. acceptor concentration here) and thickness of p-layer affect the overall performance of HIT (heterojunction with thin intrinsic layer) p-a-Si:H/i-a-Si:H/(n)c-Si solar cells. Doping of p-layer is increased from 1×10^{19} cm^{-3} to 5×10^{20} cm^{-3}, and thickness is varied from 5 to 15 nm. The best device performance is achieved at acceptor concentration of 1×10^{20} cm^{-3} and 5 nm thin p-layer. Solar cell parameters of champion device are current density (J_{sc}) of 38.22 mAcm^{-2}, open circuit voltage (V_{oc}) of 689.60 mV, fill factor (FF) of 0.83 and conversion efficiency (η) of 22.06%.

Keywords AFORS-HET · HIT solar cell · Solar cell parameters

1 Introduction

Silicon heterojunction solar (SHJ) cells are leading nowadays in the photovoltaic research areas because of its low temperature fabrication method and possibilities of achieving higher efficiencies [1, 2]. Lots of improvisation are going on to fabricate efficient solar cells. These include surface passivation to reduce recombination losses, silicon wafer thinning to reduce carrier collection losses and band gap tuning to reduce parasitic absorption losses. This requires optimization of wafer thickness, film quality and properties of doped and intrinsic hydrogenated amorphous silicon layers. Simulation studies help us to tune the properties like thickness, band gap and dopant concentration of absorber, and intrinsic and doped amorphous layer and optimize these properties for fabricating efficient SHJ solar cells [3].

J. Kumari
School of Energy Science and Engineering, Indian Institute of Technology Guwahati, Guwahati, Assam 781039, India

Rahul · P. Agarwal (✉)
Department of Physics, Indian Institute of Technology Guwahati, Guwahati, Assam 781039, India
e-mail: pratima@iitg.ac.in

V. S. Moholkar et al. (eds.), *Sustainable Energy Generation and Storage*,
https://doi.org/10.1007/978-981-99-2088-4_10

Here, we have carried out simulation studies using AFORS-HET software to understand how the performance of p-a-Si:H/i-a-Si:H/(n)c-Si photovoltaic device is affected by acceptor concentration and thickness of p-layer.

2 Simulation Tool Details

For modelling the device structure, well-known user-friendly tool AFORS-HET is used. A. Froitzheim et al. group from Japan have developed this modelling tool and made it available for all research communities in 2003 [4]. This tool offers designing of both homojunction as well as heterojunction photovoltaic devices. AFORS-HET software uses 1D equation based on steady-state conditions. As this software is widely used for modelling and analysing silicon-based solar cells, amorphous silicon (a-Si:H) and standard crystalline silicon (c-Si) layer properties are already provided in this tool. a-Si:H contains lots of defects which can be either of donor type or acceptor type [5]. All types of possible defects are taken into consideration for designing a-Si:H layer. Back and front electrodes contact potential effect are ignored by taking into account flat band conditions. Current density and voltage (J-V) characteristics, external quantum efficiency (EQE) and internal quantum efficiency (IQE) of designed devices can be easily obtained using this user-friendly tool. The standard illumination condition of AM 1.5 and power density of 100 $mWcm^{-2}$ at room temperature is used for obtaining J-V characteristics. The device configuration considered to carry out simulation studies is TCO/p-a-Si:H/i-a-Si:H/(n)c-Si/Ag. Table 1 contains details of each layer's properties used for modelling HIT solar cells [6]. The influence of acceptor concentration and thickness of p type a-Si:H layer on the performance of p-a-Si:H/i-a-Si:H/(n)c-Si HIT solar cells are studied. The acceptor concentration of p-a-Si:H is increased from 1×10^{19}, 5×10^{19}, 2×10^{20} to 5×10^{20} cm^{-3} and p-a-Si:H layer thickness is tuned from 5, 7, 10 to 15 nm keeping intrinsic layer thickness fixed at 5 nm.

3 Results and Discussion

3.1 *Effect of p-Layer Doping*

Firstly, p-a-Si:H layer doping effect on performance of p-a-Si:H/i-a-Si:H/(n)c-Si solar cell is investigated. The acceptor concentration of p-layer is tuned from 1×10^{19} cm^{-3} up to 5×10^{20} cm^{-3}. Here, thickness of p-layer(p-a-Si:H), i-layer(i-a-Si:H) and absorber layer((n)c-Si) is kept constant at 7 nm, 5 nm and 300 μm, respectively. The minority carrier lifetime is maintained at $\tau = 100$ μs. Energy band alignment and J-V characteristics of simulated devices are presented in Fig. 1a, b, respectively. From the J-V characteristics Fig. 1b, it is concluded that V_{oc} and J_{sc} have decreased

Table 1 Layer properties used for modelling p-a-Si:H/i-a-Si:H/(n)c-Si HIT solar cell

Properties and units, h for holes and e for electrons	(p)a-Si:H	(i)a-Si:H	(n)c-Si
Thickness (nm)	Variable	5 nm	300 μm
Acceptor concentration (N_A) (cm^{-3})	Variable	0	0
Donor concentration (N_D) (cm^{-3})	0	0	1^{16}
Bandgap (E_g) (eV)	1.84	1.84	1.17
Electron affinity (χ) (eV)	3.90	4.00	4.05
Holes mobility (μ_h) (cm^2V^{-1} s^{-1})	5	5	424.60
Electrons mobility (μ_n) (cm^2V^{-1} s^{-1})	20	20	1107
Layer density (ρ) (gcm^{-3})	2.32	2.32	2.32
Effective density states for valence band (N_V) (cm^{-3})	1×10^{20}	1×10^{20}	2.68×10^{19}
Effective density states for conduction band (N_C) (cm^{-3})	1×10^{20}	1×10^{20}	2.84×10^{19}
Dielectric constant (unitless)	11.90	11.90	11.90
Holes thermal velocity (v_h) (cms^{-1})	1×10^7	1×10^7	1×10^7
Thermal velocity of electrons (v_e) (cms^{-1})	1×10^7	1×10^7	1×10^7
Hole recombination (Auger) coefficient (ra_h) (cm^6s^{-1})	0	0	9.9×10^{-31}
Electron recombination (Auger) coefficient (ra_e) (cm^6s^{-1})	0	0	2.2×10^{-31}

when doping is increased from 1×10^{19} cm^{-3} to 5×10^{19} cm^{-3} but these two remain unchanged for further increase in the acceptor concentration. From the simulation results, decrease in open circuit voltage from 696.50 to 689.60 mV is observed as the doping concentration increased from 1×10^{19} cm^{-3} to 5×10^{19} cm^{-3} and it remained constant for further increase in the doping up to 5×10^{20} cm^{-3}. It may be because of increase in the minority carrier recombination rate as the doping is increased to 5×10^{19} cm^{-3}. Decrease in short circuit current is observed with increase in the acceptor concentration. Carrier–carrier scattering and ionized impurity scattering have heightened when acceptor concentration (N_A) is increased, this may reduce hole/electron mobility, and thus, recombination probability increases [7]. However, an increase in FF is observed with the increasing N_A concentration, and this is because of efficient transport of majority carriers across the interface. Parameters of p-a-Si:H/i-a-Si:H/(n)c-Si HIT solar cells are tabulated in Table 2.

It can be clearly seen from energy band alignment presented in Fig. 1a, that acceptor concentration of 1×10^{19} cm^{-3}, supports efficient electron transportation from p-a-Si:H layer to (n)c-Si. However, hole transportation from (n)c-Si to p-a-Si:H is hindered because of high valence band offset (E_{vp}–E_{vn}). Thus, efficient transportation of holes does not take place across interface causing the increased recombination at interface, resulting in lower fill factor. When the acceptor concentration is changed to 5×10^{19} cm^{-3}, fill factor has improved significantly, i.e. from 0.55 to 0.83. The reason of this improvement can be explained by the energy band alignment presented in Fig. 1a. This shows decrease in valence band offset, which supports easy holes' movement across p-a-Si:H /(n)c-Si interface and thus improved carriers' extraction from the device. That is why, the overall performance of device

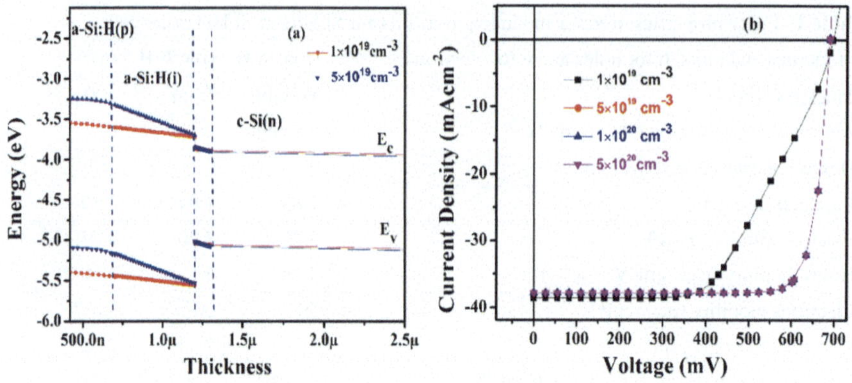

Fig. 1 **a** Energy band alignment and **b** J-V plots of simulated devices

Table 2 Cell parameters of simulated p-a-Si:H/i-a-Si:H/(n)c-Si devices

N_A (cm^{-3})	V_{oc} (mV)	J_{sc} (mAcm^{-2})	FF	η (%)
1×10^{20}	696.50	38.59	0.55	15.03
5×10^{20}	689.60	37.79	0.83	21.72
1×10^{20}	689.60	37.64	0.83	21.71
5×10^{20}	689.60	37.56	0.83	21.69

has enhanced effectively for 5×10^{19} cm^{-3} acceptor concentration. Figure 2a, b shows influence of p-layer doping on FF and power conversion efficiency (η), V_{oc} and J_{sc}, respectively. However, the overall performance of designed device has saturated for acceptor concentration $\geq 5\times10^{19}$ cm^{-3}. The acceptor concentration of 1×10^{20} cm^{-3} is considered for the next study. The best cell parameters achieved are $V_{oc} = 696.60$ mV, $J_{sc} = 37.79$ mAcm^{-2}, FF = 0.83 and $\eta = 21.71\%$.

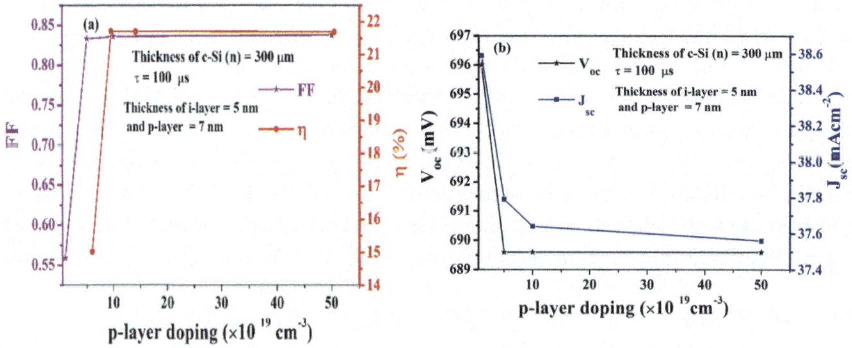

Fig. 2 Variation of **a** fill factor (FF) and conversion efficiency (η), **b** open circuit voltage (V_{oc}) and short circuit current density (J_{sc}) with p-layer doping

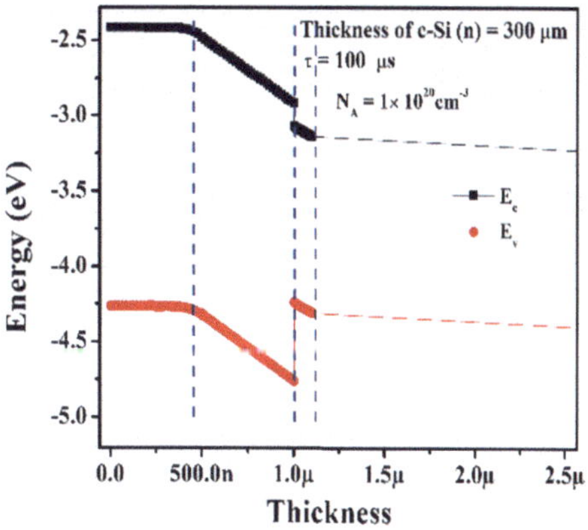

Fig. 3 Energy band alignment of simulated p-a-Si:H/i-a-Si:H/(n)c-Si solar cells

3.2 Effect of p-Layer Thickness

The optimized conditions of the previous section are used for modelling HIT silicon solar cells. Here, keeping 5 nm i-layer thickness and acceptor concentration 1×10^{20} cm^{-3}, p-a-Si:H layer thickness is increased from 5 to 15 nm. Figure 3 shows energy band alignment of simulated HIT solar cells. Quantum efficiency (QE) is the probability of incident photon's contribution towards generation and collections of photo-generated electrons [8]. EQE is one of the types of QE. It is the ratio of short circuit current a solar cell generates to incident photon flux at each wavelength. So, it mainly gives information about the optical response of a device. As the photocurrent depends on the absorption of photons and generation of carriers,

EQE can be used to describe changes taking place in J_{sc} values. Figure 4a represents J-V characteristics for varied p-layer thickness. These plots show that V_{oc} remains nearly constant whereas J_{sc} of simulated devices decreased when thickness of p-layer is increased. Thicker p-layer absorbs more photons causing parasitic absorption losses in doped silicon layer. These losses result in decreasing photons fraction which reaches to absorber layer ((n)c-Si). This decreases the photo-generated carriers' concentration and thus decreases photocurrent. It can also be justified from EQE curve presented in Fig. 4b. EQE spectrum shows that p-a-Si:H layer having thickness of 15 nm has highest photon absorption in lower wavelength region, whereas reduction in photon absorption is observed when thickness is decreased from 15 to 5 nm, resulting in increase in J_{sc} values. Figure 5a shows p-a-Si:H layer thickness effect on FF, η, and Fig. 5b presents thickness variation effect on V_{oc} and J_{sc}. From these studies, we have optimized that N_A of 1×10^{20} cm^{-3}, p-layer of thickness 5 nm and i-layer of thickness 5 nm have resulted in the most efficient solar cell having η of 22.06%, FF of 0.83, J_{sc} of 38.22 mAcm^{-2} and V_{oc} of 689.60 mV. Cell parameters obtained from simulation of p-a-Si:H/i-a-Si:H/(n)c-Si solar cells are mentioned in Table 3.

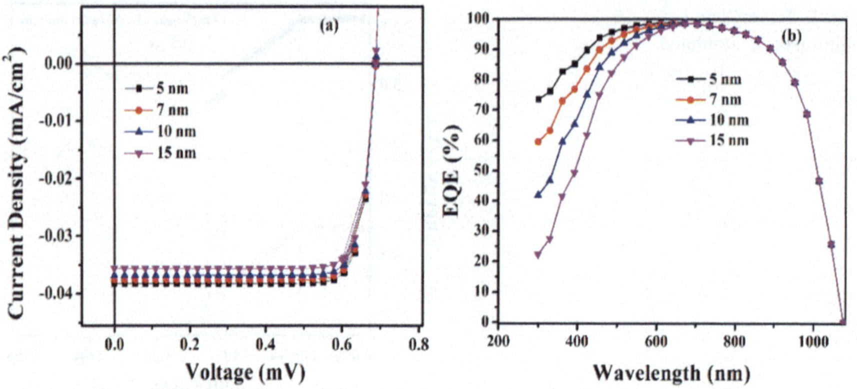

Fig. 4 **a** J-V curve and **b** EQE spectra of simulated p-a-Si:H/i-a-Si:H/(n)c-Si devices for varied p-a-Si:H layer thickness

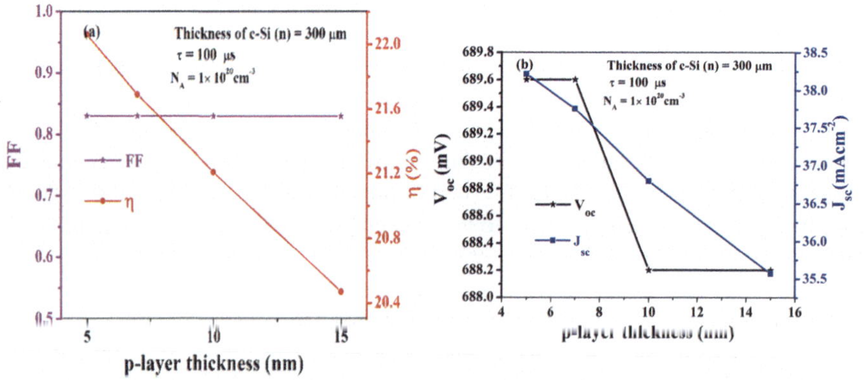

Fig. 5 **a** FF, η and **b** V_{oc}, J_{sc} of simulated device for different p-a-Si:H layer thickness

Table 3 Simulated p-a-Si:H/i-a-Si:H/(n)c-Si solar cells parameters

Thickness of p-a-S:H layer (nm)	J_{sc} (mAcm^{-2})	V_{oc} (mV)	FF	η (%)
5	38.22	689.60	0.83	22.06
7	37.76	689.60	0.83	21.69
10	36.80	688.20	0.83	21.21
15	35.57	688.20	0.83	20.47

4 Conclusion

Here, we concluded that best conversion efficiency of 22.06%, FF of 0.83, J_{sc} 38.22 mAcm^{-2} and V_{oc} of 689.60 mV are obtained for 1×10^{20} cm^{-3} N_A

concentration, 5 nm thin p-layer thickness and 5 nm i-layer thickness. These studies show that doped a-Si:H layer dopant concentration (N_A) and thickness, affect different solar cell parameters and thus have noteworthy effect on the overall performance of solar cells.

For further improvement in device performance, optimisation of other parameters like interfacial defects band gap and work function of absorber layer and n type or p type doped a-Si: H layer are required.

Acknowledgements We confess our gratefulness for providing all the requirements needed to carry out the research work to School of Energy Science and Engineering and Department of Physics, IIT Guwahati, Assam. We also acknowledge MHRD for providing funds.

References

1. V. Kanneboina, R. Madaka, P. Agarwal, High open circuit voltage c-Si/a–Si: H heterojunction solar cells: influence of hydrogen plasma treatment studied by spectroscopic ellipsometry. Sol. Energy **166**, 255–266 (2018)
2. K. Yoshikawa et al., Exceeding conversion efficiency of 26% by heterojunction interdigitated back contact solar cell with thin film Si technology. Sol. Energy Mater. Sol. Cells **173**, 37–42 (2017)
3. H. Deka, A. Sunaniya, P. Agarwal, Design and simulation of highly efficient one-sided short PIN diode silicon heterojunction solar cell. IEEE J. Photovoltaics **12**(1), 204–212 (2022)
4. A. Froitzheim, R. Stangl, L. Elmer, M. Kriegel, W. Fuhs, AFORS-HET a computer-program for the simulation of heterojunction solar cells to be distributed for public use.pdf. in *3rd World Conference on Photovolfaic Energv Conversion, May 11–18, 2003* (Osokn. Jopon, 2003)
5. W. Lisheng, C. Fengxiang, A. Yu, Simulation of high efficiency heterojunction solar cells with AFORS-HET. J. Phys. Conf. Ser. **276** (2011)
6. N. Selmane, A.C. Hikmat, S. Hilal, A numerical simulation to achieve a high efficiency influence and optimization of thickness and doping concentration of emitter layer of heterojunction solarcells. RSSI J. Sci. Eng. Sci. **09**(02) (2020)
7. S. Zhong, X. Hua, W. Shen, Simulation of high-efficiency crystalline silicon solar cells with homo-hetero junctions. IEEE Trans. Electron Devices **60**(7), 2104–2110 (2013)
8. S.C. Singh, Solar photovoltaics fundamentals—technologies and applications (2012)

Analysis of Food Waste as Potential Substrate for Biohydrogen Production

Avinash Anand and Vijayanand S. Moholkar

Abstract Biohydrogen is a clean and carbon-free fuel. It is a renewable and transitory by-product of various microbial-driven biochemical reactions. Biohydrogen generation from diverse organic waste is more feasible since it satisfies the waste reduction and energy production goals. In this present research study, enzymatic hydrolysis of food waste and effect of glucose concentration on biohydrogen production was studied. In a batch reactor, food waste hydrolyzate was employed as a substrate for biohydrogen synthesis. Solid-to-liquid ratio of 7% w/v food waste was taken as substrate for biohydrogen production, and the maximum cumulative hydrogen generation of 97 ml was attained in 21 h. It was observed that biomass quantity increased during hydrogen production process, which showed that H_2 is a growth-associated product. Some measurable metabolites (butyrate, lactate, succinate, and acetate) were also analyzed by HPLC techniques, which formed during H_2 production process.

Keywords Biohydrogen · Food waste · Enzymatic hydrolysis · Batch reactor · Metabolites · Acids

1 Introduction

1.1 Biohydrogen Production

Concerned about climate change, dwindling petroleum reserves are fueling resurgence in the search of alternatives, viz., renewable fuels. Hydrogen is one such possible candidate, which is also regarded as the cleanest fuel with only water as the combustion product. The majority of hydrogen is now produced from fossil fuels (e.g., steam reforming of natural gas). Although prototype hydrogen vehicles have

A. Anand · V. S. Moholkar (✉)
Department of Chemical Engineering, Indian Institute of Technology Guwahati, Guwahati, Assam 781039, India
e-mail: vmoholkar@iitg.ac.in

V. S. Moholkar et al. (eds.), *Sustainable Energy Generation and Storage*,
https://doi.org/10.1007/978-981-99-2088-4_11

been built, there is presently no large infrastructure for delivering hydrogen as a transportation fuel, and in-vehicle storage capacity remains a problem. Furthermore, hydrogen fuel cells are cost prohibitive, fragile, and have a short service life. Biohydrogen generation from carbohydrate-rich organic wastes and industrial wastewaters has been identified as one of the most promising and sustainable alternatives to fossil fuels, as well as a means of mitigating the greenhouse effect [1]. Biological processes such as direct and indirect biophotolysis, dark fermentation, and photofermentation can all generate hydrogen. Dark fermentation is a significant approach among them because of the faster rate of hydrogen evolution and the diversity of the organic substrates utilized. However, the main limitation of this method is poor yield of hydrogen per mole of organic substrate [2]. Biohydrogen is known as the smallest biological substrate and has immense capability as a fossil fuel substitute. It generates solely water as a by-product of combustion, making it a non-polluting energy source. It has high-energy content (or heat of combustion) of 142 kJ/g, which is approximately three times more than that of petrol or diesel [1]. Biohydrogen (H_2) is produced by microorganisms in two ways: photosynthesis and fermentation. Fermentation processes generally give higher magnitude of hydrogen than photosynthetic process, do not require light, use different varieties of renewable resources, use less energy, and are technically much simpler and more stable, which appear to have more potential for practical applications [3]. Many waste materials from agriculture sector, diary sector, and other lignocellulosic wastes are being converted to hydrogen, but the entire methods of pretreatment process make it a lengthy and less economic process. Among these fermentative hydrogen producers, the strict anaerobic bacteria of *Clostridium* species such as *C. butyricum* [4], *C. acetobutylicum*, *C. saccharoperbutylacetonicum* [5], and *Clostridium pasteurianum* are always used to produce H_2. The H_2 yield from glucose depends on fermentation routes and its end products. The overall *Clostridial* fermentation can be

$$C_6H_{12}O_6 + 2H_2O \longrightarrow 2CH_3COOH + 2CO_2 + 4H_2 \quad (1.a)$$

$$C_6H_{12}O_6 \longrightarrow C_4H_9COOH + CO_2 + 2H_2 \quad (1.b)$$

Pure or mixed facultative and stringent anaerobe cultures can carry out dark fermentation. *Clostridium thermocellum, Clostridium acetobutylicum, and Clostridium pasteurianum* are more efficient hydrogen producers than Bacillus spp., *E. coli*, and *Enterobacter* spp. The metabolic pathway for hydrogen production differs to some extent in both types of anaerobes. The (Fig. 1) pyruvate formate lyase (PFL) or the pyruvate ferredoxin oxidoreductase (PFOR) pathways convert sugars to pyruvate, acetyl-coA, formate or reduced ferredoxin, and CO_2, depending on the organism. Formate hydrogenase, which usually comprises [NiFe] hydrogenase, converts formate to hydrogen and CO_2. Only PFOR species may use NADH produced during glycolysis to create hydrogen through NADH-dependent [Fe–Fe] hydrogenase. Another class of [Fe–Fe] hydrogenase produces hydrogen by re-oxidizing ferredoxin and is called Fd-dependent [Fe–Fe] hydrogenase.

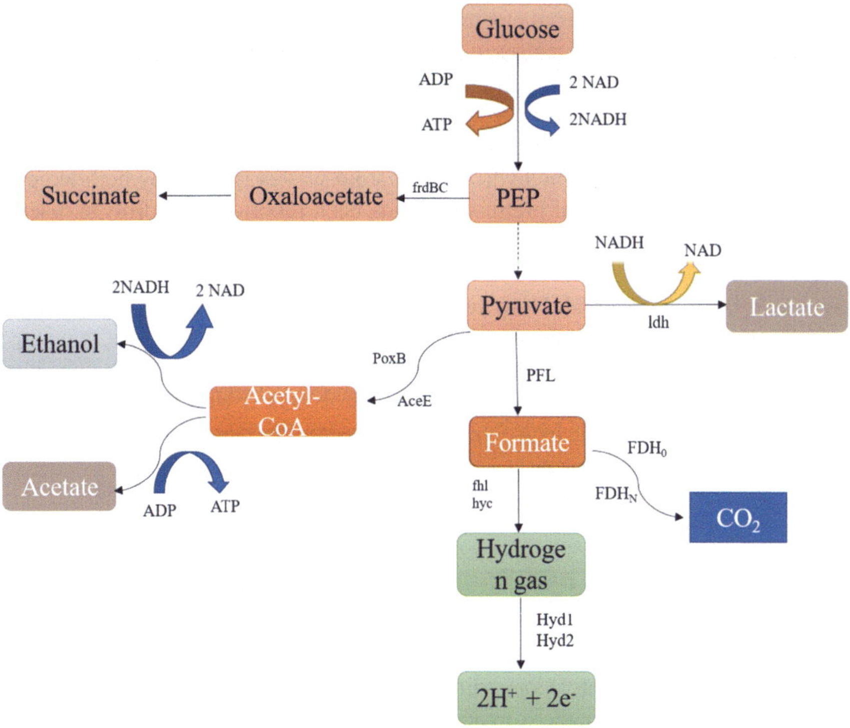

Fig. 1 Metabolic pathway for hydrogen production for facultative anaerobes via pyruvate formate lyase (PFL pathway) [6]

1.2 *Food Waste as Substrate for Hydrogen*

Currently, major route for hydrogen production is based on fossil fuels [7]. But due to the limited fossil fuel reserves, quest for alternative renewable sources/substrates for hydrogen production is on. Food waste is one such alternative resource for hydrogen production. Simple carbohydrates like glucose and sucrose are excellent substrates for the generation of biological hydrogen. Hydrogen generation from mixed organic waste is more feasible, since it achieves both waste reduction and energy production goals. Because of its high quantity of organic matter and carbohydrates, as well as its highly hydrolyzable and biodegradable nature, food waste (FW) has a significant hydrogen generation potential [8]. Uneaten food and food preparation leftovers from homes, commercial venues such as restaurants, institutions such as school cafeterias, and industrial sources such as factory lunch rooms are all examples of food waste [9]. Despite diverse sources of food waste, it mainly comprises carbohydrates, protein, and fat, with minor quantities of cellulose and hemicellulose. The ratio of volatile solids to total solids (VS/TS) is greater than 80%. These volatile solids degrade quickly and can be converted to biogas in 80–90% of cases [10]. The probability of producing biohydrogen from enzymatic hydrolysis of food waste was studied in this study. The rate-limiting phase in the whole anaerobic fermentation process is hydrolysis of food waste [14]. Different pretreatment techniques, such as acidic or basic hydrolysis, have been proposed to improve hydrogen generation. Autoclaving, which includes steam processing under pressure, was recently adopted for municipal solid waste treatment. The hydrolysis of solid wastes is aided by high temperature and pressure. Chemical pretreatment has been reported to hydrolyze macromolecules into micromolecules. Enzymatic hydrolysis, which can convert starch in food waste into glucose under ordinary conditions and without producing inhibitors, might be a potential method. However, there is a scarcity of knowledge on biohydrogen synthesis via enzymatic hydrolysis of food waste. We created a two-stage bioprocess for producing hydrogen from food waste to overcome these concerns. The food waste hydrolyzate was created by hydrolyzing it with the glucoamylase enzyme first. The food waste hydrolyzate was then employed as a substrate in a batch reactor to produce biohydrogen.

1.3 Aim, Utilization, and Scope of Present Study

The goal of our research is to employ glucose, food waste hydrolyzate, as a biohydrogen production substrate. Converting these substrates to a clean fuel is a financially viable idea. However, there are just a few studies in the prior literature that have addressed this topic. *C. pasteurianum*, as well as mixed cultures of *C. pasteurianum* and *C. acetobutylicum*, is gram-positive bacteria that can use glucose as a source of energy and convert it to hydrogen and other metabolites (Fig. 2).

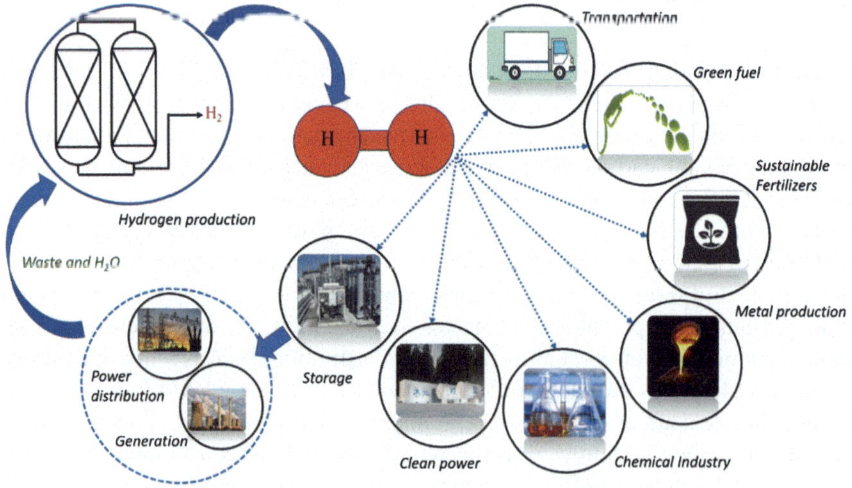

Fig. 2 Benefits of hydrogen economy

2 Materials and Methods

2.1 Feedstock and Materials

Food waste (FW) sources were collected from IIT Guwahati canteen/mess. Food waste contained mixed sugars and carbohydrates. FW was mainly food remaining in plates after lunch consisted of boiled rice, dal, and fried or boiled potatoes. Food waste was dried for 48 h in oven at 60 °C. After drying, it was grinded into powdered form. Commercial glucoamylase purchased from Rich Core Life Sciences (P) Ltd. was used for enzymatic hydrolysis of food waste. The activity of glucoamylase was given to be 1,35,000–1,95,000 U/ml by the supplier. *C. pasteurianum* (ATCC 6013) MTCC 116 microbial culture was purchased from the (MTCC), Chandigarh, India. All other RCM/RCB media components and chemicals were used in this study provided by Sigma-Aldrich and HiMedia Pvt. Ltd., India, [9, 11].

2.2 Enzymatic Hydrolysis of Food Waste

Commercial glucoamylase enzyme from Rich Core Life Sciences (P) Ltd. was purchased to perform enzymatic hydrolysis of food waste. Enzymatic hydrolysis was carried out in an incubator shaker at 55 °C and 200 rpm in a 50 mM citrate phosphate buffer solution (pH 4.5). The reaction mixture was placed in a 500 mL glass flask with a 100 mL total reaction volume. The pretreatment biomass content in the reaction mixture was 7% w/v, with a glucoamylase concentration of 175 FPU/mL biomass. The hydrolysis process was continued till 120 h. During enzymatic hydrolysis, 0.5 mL samples of the reaction mixture were taken and analyzed for the release of reducing sugar. As glucose concentration in reaction mixture reached saturation, hydrolysis was terminated. To get the liquid food waste hydrolyzate, the centrifugation of mixture (food waste hydrolyzate) was done at 10,000 rpm for 20 min before filtration with filter paper. This hydrolyzate was used as a biohydrogen production substrate. Notably, residual oil traces in food waste were also removed by this treatment.

2.3 Batch Fermentation

The cells of *C. pasteurianum* were grown up in fresh reinforced clostridia media (RCM)/reinforced clostridia broth (RCB) before being inoculated under oxygen-free conditions or anaerobic environment. Batch studies were conducted by transferring 10% v/v inoculum from the mid-log phase of culture into a 100 mL serum bottle with RCM/RCB medium with a working volume of 50 mL, which included 5 mL (10% v/v) *C. pasteurianum* inoculum and 0.5 mL of 6% cysteine hydrogen chloride

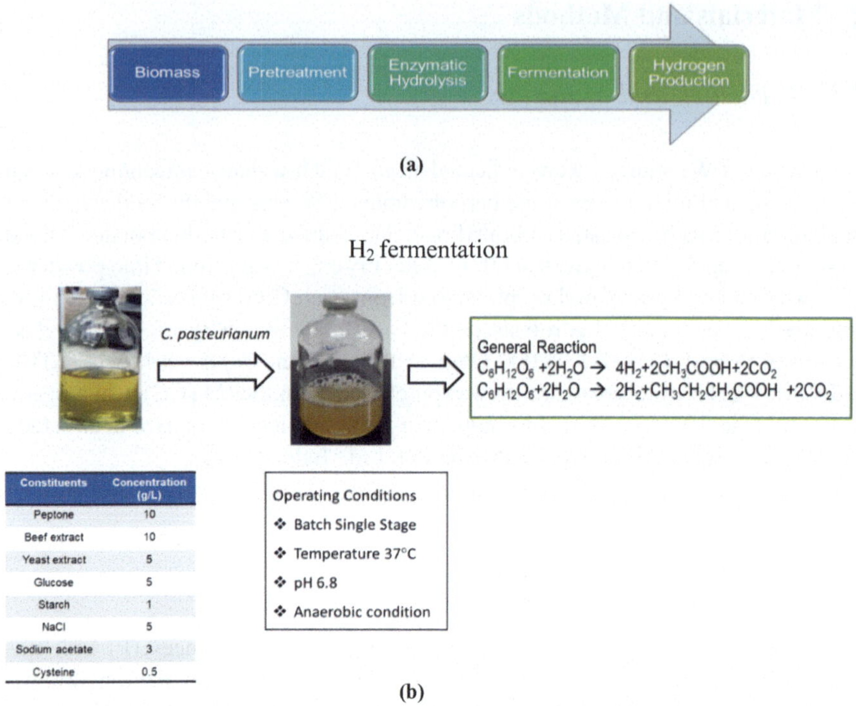

Fig. 3 **a** Stepwise methods of biohydrogen production, **b** H_2 fermentation process

(w/v) as a reducing agent. The RCM included peptone (10), yeast extract (5), beef extract (10), glucose (5), starch (1), sodium chloride (5), and sodium acetate (3) (concentration, g/L). About 1 M NaOH was used to bring the pH of the medium down to 7. The medium was purged or flushed with (99.99%) pure nitrogen gas for 20 min prior to inoculation, then serum bottles were closed up with rubber stoppers, crimped, and autoclaved for 15 min at 121 °C temperature and 15 psi pressure. After that, the serum bottles were inoculated with a 10% v/v (5 ml) inoculum and incubated at 37 °C with 180 rpm. For reproducibility, each experiment was conducted twice. To determine how much glucose/reducing sugar was consumed, aliquots of fermentation broth were taken out at regular time intervals [12, 13]. The production of biohydrogen carried out by following steps and their fermentation process is shown in Fig. 3b.

2.4 Analytical Methods

Centrifugation was used to separate samples of the fermentation mixture at 12,000 rpm for 20 min; a membrane filter with a pore size of 0.25 m was used to filter the supernatant. A refractive index detector (87 H) column, a pump, and a vacuum degasser formed the HPLC system (Perkin Elmer, Series 200). HPLC was used to evaluate the concentration of mixed reducing sugars and some important metabolites in the sample using a monosaccharide calcium column. Column oven temperature was set at 50 °C. The HPLC used ultra-pure water mixed with 0.5 mM concentrated sulfuric acid as the mobile phase (flow rate = 0.6 mL/min). The glucose used throughout the fermentation process was measured using standard calibration plots. To quantify and identify the composition of the product gas collected, a gas chromatograph was used. The gas chromatogram (GC) setup includes a thermal conductivity detector (TCD) set to 200 °C, a Porapak Q (60/80 mesh) GC column in which column oven temperature was at 50 °C, and injector temperature was at 200 °C, respectively. Argon (Ar) gas was used as carrier gas in gas chromatogram (GC) (flow rate: 35 mL/min). The concentration of H_2 and CO_2 was determined by injecting gaseous samples from the headspace of serum bottles. With the help of given mass balance equation, total volume of gas can be calculated [12, 14].

$$V_{H,j} = V_{H,j-1} + C_{H,j}\left(V_{G,j} - V_{G,j-1}\right) + V_H\left(C_{H,j} - C_{H,j-1}\right) \quad (1.c)$$

V_H denotes volume of headspace in serum bottles. $C_{H,j}$ and $C_{H,j-1}$ indicate volume or mole fractions of hydrogen gas in the headspace at the current and preceding time periods. During the hydrogen mass balance, the volume of produced hydrogen gas needed for analysis was also taken into consideration. The cumulative biohydrogen gas volumes at the current (j) and preceding (j–1) time intervals are $V_{H,j}$ and $V_{H,j-1}$, respectively.

3 Results and Discussion

Time profiles of reducing sugar in enzymatic hydrolysis of food waste are shown in Fig. 4a. It could be seen from Fig. 4a that the highest sugar yield was obtained for glucoamylase enzyme concentration of 5 μL in 100 mL reaction volume of the mixture. For all three enzyme concentrations used in this study (viz., 5, 10 and 100 μL), the reducing sugar concentration reached saturation after approx. 96 h of treatment.

3.1 *Effect of Glucose Concentration on Biohydrogen Production*

The production of biohydrogen is influenced by variation in glucose concentrations. The effect of glucose concentration was studied at three different concentrations 5, 10, and 15 g/L. It could be observed from Fig. 5b that at 5 g/L glucose, hydrogen production increases from 0.44% v/v at 3 h to 45% v/v at 18 h. Thereafter, the H_2 production

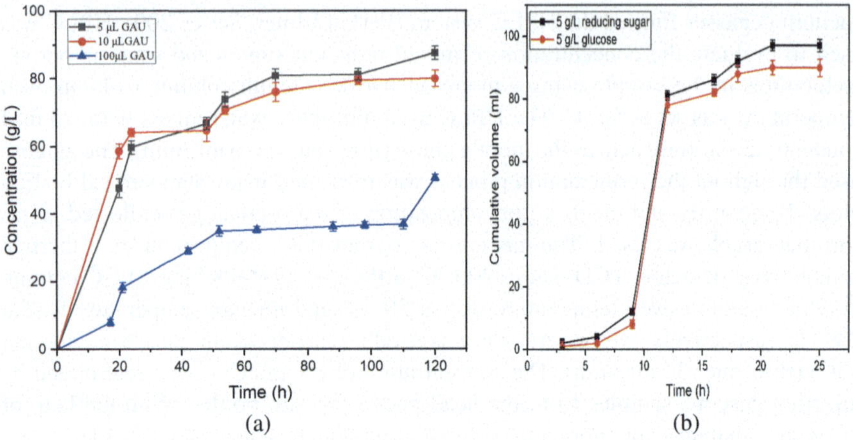

Fig. 4 **a** Enzymatic hydrolysis of food waste. **b** Comparison of cumulative H_2 production by *C. pasteurianum* at for 5 g/L pure glucose and food waste hydrolyzate at reducing sugar concentration of 5 g/L

remained constant. In case of 10 g/L glucose concentration, hydrogen production increases from 1.79% v/v at 3 h to 52% v/v at 15 h. For 15 g/L glucose concentration, and production of hydrogen increases from 3.97% v/v at 3 h to 45.35% v/v at 24 h. Similar trend was observed with cumulative volume of hydrogen with 10 and 15 g/L glucose as substrate with maximum volume obtained at 18 h for 10 g/L, which is shown in (Fig. 5a).

The biomass profile of *C. pasteurianum* was also obtained with glucose as substrate. From Fig. 5c, it could be inferred that the log phase of *C. pasteurianum* occurred between 5 and 18 h, after which the microbial culture was in its late log phase. The H_2 production was also observed to be maximum during this period, thus concluding that H_2 is a growth-associated product.

3.2 *Analysis of Metabolites*

HPLC analysis of sample shows that concentration of lactate, acetate, and butyrate is higher at 24 h as compared to concentration of same metabolites at 3 h. It also shows that concentration of succinate is approximately equal in both time periods. Higher concentration of butyrate indicates that butyric acid-associated metabolite reaction (Fig. 6).

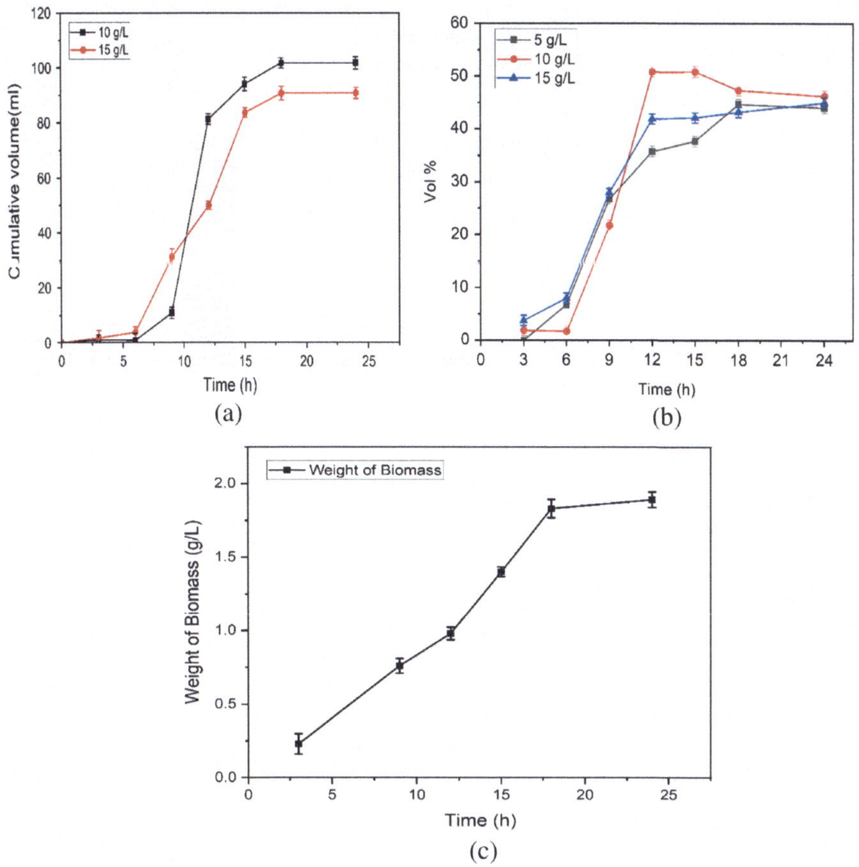

Fig. 5 **a** Time profile of cumulative hydrogen production by *C. pasteurianum* at various concentrations of glucose (10 g/L and 15 g/L) as substrate. **b** Time profile of percentage hydrogen production by *C. pasteurianum* at various concentrations of glucose (5 g/L, 10 g/L and 15 g/L) as substrate. **c** Time profile of biomass production of *C. pasteurianum* with food waste as a substrate

4 Conclusion

It is revealed that hydrogen production increased from 3 to 18 h and thereafter remained constant. From biomass profile, it was also observed that biomass production was maximum during this period, which showed that hydrogen is a growth-associated product. The hydrogen yield for food waste hydrolyzate with reducing sugar concentration of 5 g/L as substrate was higher than pure glucose substrate with same concentration. Measurable metabolites analyzed by HPLC were lactate, acetate, glucose, succinate, and butyrate. The concentration of butyrate was higher than concentration of acetate. Therefore, hydrogen production process is butyric acid-based metabolic reaction.

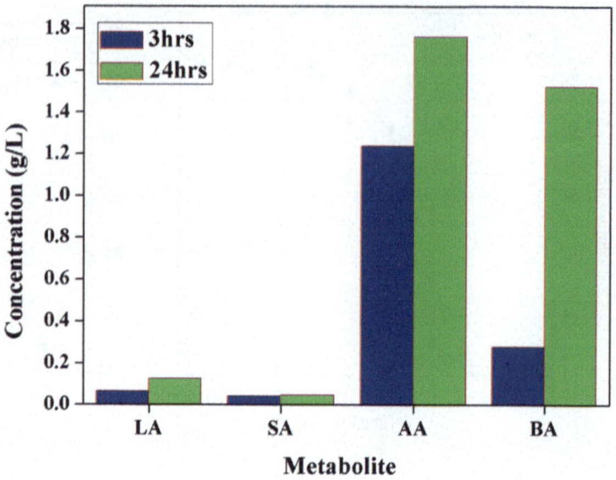

Fig. 6 Concentration profiles of metabolites in *C. pasteurianum* with glucose as fermentation substrate. Abbreviations: LA—lactate, SA—succinate, AA—acetate, BA—butyrate

References

1. A. Gadhe, S.S. Sonawane, M.N. Varma, Int. J. Hydrogen Energy **38**, 6607 (2013)
2. D. Das, Int. J. Hydrogen Energy **26**, 13 (2001)
3. G. Balachandar, J.L. Varanasi, V. Singh, H. Singh, D. Das, Int. J. Hydrogen Energy **45**, 5202 (2020)
4. I. Hussy, F. Hawkes, R. Dinsdale, D. Hawkes, Int. J. Hydrogen Energy **30**, 471 (2005)
5. N. Kumar, D. Das, Enzyme Microb. Technol. **29**, 280 (2001)
6. J. Mathews, G. Wang, Int. J. Hydrogen Energy **34**, 7404 (2009)
7. P. Nikolaidis, A. Poullikkas, Renew. Sustain. Energy Rev. **67**, 597 (2017)
8. I.K. Kapdan, F. Kargi, Enzyme Microb. Technol. **38**, 569 (2006)
9. N.H.M. Yasin, T. Mumtaz, M.A. Hassan, and N. Abd Rahman. J. Environ. Manage. **130**, 375 (2013)
10. W. Han, Y. Yan, Y. Shi, J. Gu, J. Tang, H. Zhao, Sci Rep **6**, 38395 (2016)
11. C. Gong, A. Singh, P. Singh, A. Singh, Indian J Microbiol **61**, 427 (2021)
12. S. Sarma, V.K. Dubey, V.S. Moholkar, Int. J. Hydrogen Energy **41**, 19972 (2016)
13. S. Singh, M. Agarwal, S. Sarma, A. Goyal, V.S. Moholkar, Ultrason. Sonochem. **26**, 249 (2015)
14. S. Sarma, A. Anand, V.K. Dubey, V.S. Moholkar, Biores. Technol. **242**, 169 (2017)

Characterization of Biofuel Obtained by Pyrolysis of *A. Indica*

Gaffer Ahmed and Nanda Kishore

Abstract Pyrolysis of *Azadirachta indica* was performed in a tubular batch reactor at a temperature of 600 °C and pressure of 1 bar with a heating rate of 10 °C/min under inert conditions. Maximum liquid yield of 25.574 mass % was observed from the pyrolysis experiments. The organic phase obtained had a density, viscosity and pH value of 1.053 g/ml, 0.933 mPa.s and 3.91, respectively. The calorific value of the organic phase was found to be 19.62 MJ/kg. The organic phase obtained mainly consisted of aliphatic compounds, ketones, ethers, alkyl compounds, aromatic compounds, hydrocarbons and alcohols. The biochar produced had fixed carbon as its major constituent, and it had a calorific value of 28.03 MJ/kg.

Keywords Biomass conversion · Green energy · Waste to energy · Thermochemical conversion · Biofuels

1 Introduction

With every passing year, people has become more dependent on energy. Among various sources of energy, people are mostly dependent on fossil fuels. Due to its excessive use, there is always a high demand of it, leading to price hike. In addition to it, excessive use of fossil fuels has many harmful impacts on the environment leading to global warming, depletion of ozone, acid rain, etc. According to Intergovernmental Planet on Climate Change (IPCC), the discharges from combustion of fossil fuels were found to be the chief cause of global warming. For the year 2018, 89% of the global CO_2 emissions came from fossil fuels and industries [1]. This has driven the world toward renewable energy. There are various sources of renewable energy such as solar, wind, biomass, tidal and geothermal [2]. Among these, biomass has achieved great importance in recent years because of its wide availability and can be converted to solid, liquid and gaseous fuels. Broadly classifying, there are two

G. Ahmed (✉) · N. Kishore
Department of Chemical Engineering, Indian Institute of Technology Guwahati, Guwahati, Assam, India
e-mail: gaffer@iitg.ac.in

V. S. Moholkar et al. (eds.), *Sustainable Energy Generation and Storage*,
https://doi.org/10.1007/978-981-99-2088-4_12

major routes of biomass conversion—biochemical conversion and thermochemical conversion [3]. Biochemical conversion uses microorganisms for the breakdown of biomass, whereas thermochemical uses heat and pressure. Among these pathways, thermochemical conversion has certain advantages over biochemical conversion such as lower residence time, low waste generation and high conversion efficiency. There are various thermochemical conversion processes such as combustion, pyrolysis, liquefaction and gasification. Among these, pyrolysis has been found to be more advantageous in terms of ease of handling, operating conditions and product distribution [4]. Thus, this work focuses on the utilization of a scarcely researched biomass "*Azadirachta indica*" for the production of biofuels through pyrolysis.

2 Materials and Methods

2.1 Biomass Collection and Characterization

The biomass used for this study is *A. indica* commonly known as Mahaneem in India that is an evergreen tree which is widely available in the Indian subcontinent. The tree is widely used as a medicinal plant because of its health beneficial properties. Apart from it, the tree is also used for making furniture, doors, wooden anticraft, etc., while the extracts of the trees were used to keep away insects from grains and cereals by mixing with it. Owing to its multipurpose utility, there is a high chance of generation of waste from it. Thus, the objective of this research study is to convert *A. indica* into biofuels of high energy density through pyrolysis.

Dry branches of *A. indica* are collected from the arboricultural area of IIT Guwahati where the tree is grown for ornamental purpose. The branches are chopped down and grounded in a hammer mill to small pieces. The small pieces are than pulverized in a mixer grinder followed by sieving using a 400-micron mesh to obtain a uniformly sized powdered feed. Before proceeding, it is very much necessary to check the feasibility of *A. indica* as a feedstock for the pyrolysis process. Hence, the physiochemical and fuel properties such as the moisture content, volatile matter content, ash content, fixed carbon content, elemental composition and calorific value were determined.

In order to determine the temperature for the pyrolytic conversion of *A. indica* biomass, approximately 10 mg of the biomass was analyzed using a TGA analyzer (make: M/S Netzsch, Germany; model no.: TG 209 F1 Libra) at temperature ranging between 25 and 800 °C, while the heating rate for the analysis was considered to be 10 °C/min. The TGA analysis shows the degradation of the biomass over a period of time with increasing temperature.

2.2 *Pyrolytic Conversion of* A. Indica *Biomass*

The pyrolytic conversion of dried and pulverized *A. indica* biomass is studied in a tubular batch reactor made of stainless steel. About 300 g of the grinded biomass is fed to the reactor, and the reactions were carried out at a temperature of 600 °C and pressure of 1 bar with a heating rate of 10 °C/min under inert conditions. The residence time of the reaction is considered for two hours after reaching the set temperature. The vapors produced during the reaction consisted of condensable and non-condensable gases which were passed through a condenser to condensate the condensable gases and the gaseous phase are separated using a gas–liquid separator. After completion of the reaction, the system is then allowed to cool down to room temperature to collect the products. The liquid product produced is generally termed as bio-oil which consists of two phases: organic phase and aqueous phase. The organic phase of the bio-oil is separated from the aqueous phase using dichloromethane. The gaseous product mainly consists of non-condensable gases composing CH_4, H_2, CO and CO_2 as its main constituents, whereas the solid product contains carbon as its prime constituent, and it is generally termed as biochar.

2.3 *Yield of Products*

The yields of solid, liquid and gaseous products obtained were obtained in wt%. The yield of biochar in wt% is obtained by dividing the weight of biochar by the weight of feed of the biomass. The yield of the organic phase and aqueous phase is obtained in milliliters and then converted to wt% by multiplying the volume of each phase to its corresponding density and dividing by the weight of the feed of the biomass. The yield of the gaseous phase is obtained by subtracting the yield of biochar and bio-oil from 100.

2.4 *Characterization of the Products*

The physical and fuel properties such as density, viscosity, pH value and calorific value of the organic and aqueous phase of the bio-oil were evaluated immediately after the separation of the organic phase of the bio-oil. The organic phase of the bio-oil is also characterized using FTIR analysis to determine the functional groups present in it. The instrument used for the FTIR analysis is (make: M/S Shimadzu, Japan; model no.: IRAffinity-1). The physiochemical and fuel properties of the biochar were determined similarly as done for the biomass.

3 Results and Discussion

3.1 Properties of the Biomass

The physiochemical and fuel properties of *A. indica* biomass are presented in Table 1. The proximate analysis indicated that the biomass had a high volatile matter content of 82.727 wt% on dry basis along with low contents of moisture and ash in it. The ultimate analysis indicates that the biomass has a high content of carbon (50%) and hydrogen (7.991%). This indicates that the biomass had a high H/C ratio (1.917) which indicates that the biomass can be converted to high-value liquid fuel products. Apart from this, the biomass has a low O/C ratio (0.533) and high calorific value (18.020 MJ/kg). These results indicate that the biomass is suitable for pyrolytic conversion and can serve as an excellent feedstock for producing high-grade fuels. The TGA and DTG analyses of *A. indica* biomass are depicted in Fig. 1. The TGA curve indicated that the degradation of the biomass occurred in three stages. The first stage of degradation is associated with the removal of moisture and lower molecular weight hydrocarbons which occurred till 180 °C. In the second stage, the degradation of cellulose and hemicellulose takes place which occurred in the range of 180–350 °C. The final stage is associated with the degradation of lignin, and it starts from 350 °C and continues till the end of the analysis. It was observed that approximately 25 wt% of the sample was left at the end of the analysis which may be due to the presence of inorganic matter in the sample or occurrence of secondary reactions during pyrolysis inside the chamber. Although from the DTG curve, it was obvious that the maximum degradation of the biomass occurred in the range of 340–360 °C; however, the pyrolysis temperature was considered to be 600 °C to achieve an optimum conversion of the biomass. A few relevant literatures related to the thermal degradation of lignocellulosic biomass have been discussed below. Kawale and Kishore [5] carried out thermogravimetric analysis of *Delonix regia* between 30 and 750 °C and found that the removal of moisture took place till 150 °C on dry basis, whereas the degradation of hemicellulose started at around 270–280 °C. The degradation of cellulose and lignin started from 300 °C and continued till 400 °C for cellulose and continued till 700 °C for the lignin. Thermogravimetric analysis of *Polyalthia longifolia* leaves was performed at temperatures ranging between 30 and 800 °C by Ahmed et al. [6]. Similar to other biomasses, thermogravimetric analysis showed removal of moisture from 80 to 150 °C. The degradation of cellulose and hemicellulose was observed between 200 and 390 °C. The degradation of lignin continued till 800 °C, and complete degradation of the lignin was not possible. Approximately, 25–30% of the biomass was left unconverted [6]. Thermal degradation of sawdust obtained from *Acacia* timber was studied between 25 and 900 °C. The TGA profile indicated removal of moisture and lower molecular volatile compounds. Between 200 and 500 °C, the degradation of cellulose and hemicellulose component took place, while the degradation of lignin component was continuous and had a long tail which continued till 900 °C [7]. Thus, it has been observed that similar to other

Table 1 Proximate analysis, ultimate analysis, H/C ratio, O/C ratio and calorific value of *A. indica* biomass

Proximate analysis (wt%)				Ultimate analysis (wt%)					H/C ratio	O/C ratio	Calorific value (MJ/kg)
Moisture content	Volatile matter content	Ash content	Fixed carbon content[a]	C	H	N	S	O[a]			
14.179	82.727	3.224	14.050	50	7.991	5.66	0	35.560	1.917	0.533	18.020

[a]Calculated by difference

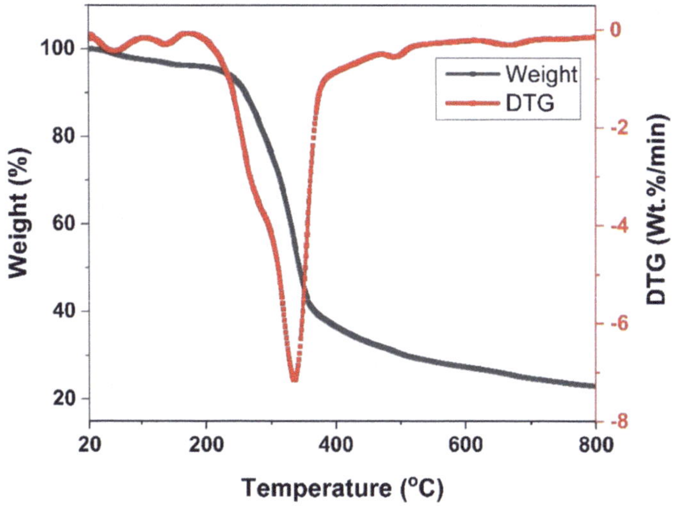

Fig. 1 TGA and DTG curve for *A. indica* biomass

lignocellulosic biomasses, *A. indica* biomass also followed similar kind of trends for TGA and DTG. The results were also in line with other literatures [8–10].

3.2 *Product Distribution*

The yields of products were obtained in wt% and are represented in Fig. 2. It was found that the maximum bio-oil yield was found to be 25.574 wt% which comprised 3.763 wt% of organic phase and 21.811 wt% of aqueous phase. The yield of biochar and non-condensable gases was found to be 40.481 and 33.945 wt%, respectively. A few relevant literatures related to pyrolysis of lignocellulosic biomass have been discussed below. Pyrolytic conversion of *Delonix regia* carried out by Kawale and Kishore [5] in the temperature range of 500–700 °C resulted in a bio-oil yield of 31.5–37.8 wt%, while the yield of biochar varied in the range of 34.7–30.7 wt%. Ahmed

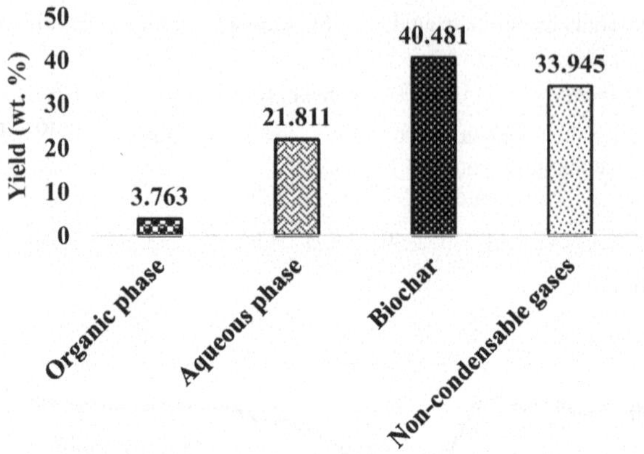

Fig. 2 Pyrolysis product distribution of *A. indica* biomass

et al. [6] carried out pyrolysis of *Polyalthia longifolia* leaves in a tubular at a temperature range of 450–600 °C. The average yield of bio-oil varied between 14.35 and 23.91 wt%, whereas the yield of biochar varied between 44.0 and 59.79 wt%. With increasing temperature, the yield of bio-oil and non-condensable gases increased, whereas the yield of biochar decreased. Another study on pyrolytic conversion of *Acacia* timber was performed in an auger reactor at temperature ranging between 400 and 600 °C. Maximum bio-oil yield was observed at 500 °C which indicated a yield of 45.1 wt%. The yield of biochar decreased with increasing temperature, while the yield of non-condensable gases increased with increasing temperature. The yield of bio-oil increased and then decreased [7]. Pyrolysis of *A. indica* biomass also followed similar trends in the yields of bio-oil, biochar and non-condensable gases.

3.3 Characterization of the Products

The physical and fuel properties such as density, viscosity, pH value and calorific value of the organic and aqueous phase of the bio-oil are presented in Table 2. It was observed that the organic phase and the aqueous phase had a density of 1.053 and 1.002 g/ml, while the viscosity was observed to be 0.933 and 1.24 mPa.s, respectively. The pH value of the organic and aqueous phase was observed to be 3.91 and 3.81, respectively. It was observed that the organic phase had a calorific value of 19.62 MJ/kg, while the aqueous phase did not have any calorific value as water is the main constituent of the aqueous phase.

The FTIR spectrum of the organic phase obtained from pyrolytic conversion of *A. indica* showed intensities at wavenumbers of cm^{-1} 3403, 2960, 2854, 1707, 1607, 1507, 1457, 1367, 1272, 1222, 1108, 1026, 750, 595, 528 and 469. A graphical representation of the same has been presented in Fig. 3. The functional groups identified in the organic phase sample have been presented in Table 3. The broad peaks appearing

Table 2 Physical properties and calorific value of the organic phase and aqueous phase of *A. indica* biomass

Organic phase				Aqueous phase		
Density (g/ml)	pH	Viscosity (mPa.s)	HHV (MJ/kg)	Density (g/ml)	pH	Viscosity (mPa.s)
1.053	3.91	0.933	19.62	1.002	3.81	1.24

between 3420 and 3250 cm^{-1} indicated the occurrence of polymeric OH and water impurities. The peaks appearing between 2990 and 2850 cm^{-1} indicate the existence of methyl and methylene groups falling under the class of aliphatic compounds. The presence of ketones was indicated by the peaks occurring in the range of 1720–1700 cm^{-1}, while the benzene ring in aromatic compounds is indicated by the peaks at 1615–1590 and 1515–1485 cm^{-1}. CH_2 group of aliphatic groups was confirmed by the peaks occurring at 1475–1450 cm^{-1}, while *t*-butyl and isopropyl groups were confirmed by the peaks at 1400–1370 cm^{-1} and 1380–1360 cm^{-1}, respectively. The peaks at 1280–1180 cm^{-1} and 1150–1085 cm^{-1} confirmed the presence of aromatic and aliphatic ethers, whereas the peaks at 1065–1015 cm^{-1} confirmed the presence of cyclic alcohols. Then, o-benzenes were confirmed by the peaks at 760–740 cm^{-1}. The peaks in the range of 700–590 cm^{-1} and 565–540 cm^{-1} indicate the existence of $O–C = O$ in carboxylic acids and $S–C \equiv N$ in thiocyanites and C_nH_{2n+1} in alkyl groups, respectively.

The biochar obtained through pyrolytic conversion of *A. indica* biomass has been characterized for the physiochemical and fuel properties and is represented in Table 4. The proximate analysis indicated that the biochar had a high fixed carbon

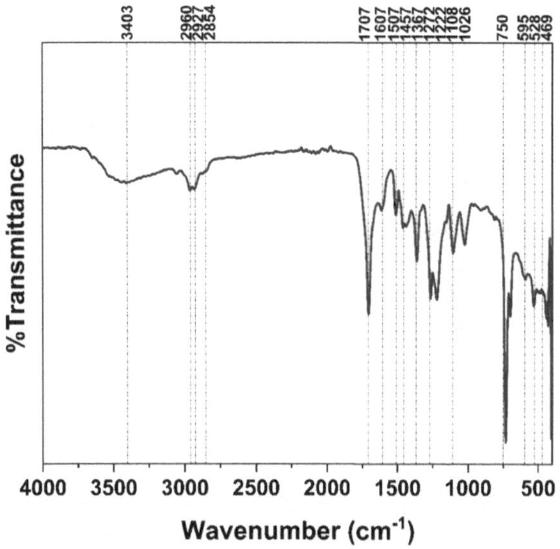

Fig. 3 FTIR spectrum of organic phase obtained from pyrolysis of *A. indica*

Table 3 Functional groups present in the organic phase of bio-oil of *A. indica*

Wavenumber (cm^{-1})	Wavenumber range (cm^{-1})	Class of compound
3403	3420–3250	Polymeric OH and water impurities
2960, 2927, 2854	2990–2850	Methyl and methylene groups of aliphatic compounds
1707	1720–1700	C = O group of ketones
1607	1615–1590	Benzene ring in aromatic compounds
1507	1515–1485	Benzene ring in aromatic compounds
1457	1475–1450	Methylene group of aliphatic compounds
1367	1400–1370	*t*-butyl group of *t*-butyl compounds
1272, 1222	1280–1180	C–H group in aromatic compounds
1108	1150–1085	Aliphatic ethers
1026	1065–1015	C–OH groups of cyclic alcohols
750	760–740	o-benzenes of aromatics
595	700–590	O–C = O in carboxylic acids
528, 469	565–440	C_nH_{2n+1} in alkyl groups

content (70.193 wt%) along with low contents of volatile (17.811 wt%) and ash content (11.996 wt%). The ultimate analysis indicated that carbon (66.568 wt%) was found to be the prime constituent followed by oxygen (29.751 wt%) and hydrogen (3.681 wt%). The low H/C ratio (0.663) indicated a major devolatilization of the biomass. The calorific value of the biochar was found to be 28.03 MJ/kg. The physiochemical and fuel properties of the biochar indicate that the biochar has properties similar to that of conventional coal and can be used as a feedstock for energy production. The other possible uses of biochar are as application as soil conditioner and bioremediation.

Table 4 Proximate analysis, ultimate analysis, H/C ratio, O/C ratio and calorific value of *A. indica* biochar

Proximate analysis (wt%)				Ultimate analysis (wt%)					H/C ratio	O/C ratio	Calorific value (MJ/kg)
Moisture content	Volatile matter content	Ash content	Fixed carbon content[a]	C	H	N	S	O[a]			
3.45	17.811	11.996	70.193	66.568	3.681	-	-	29.75	0.663	0.335	28.03

[a]Calculated by difference

4 Conclusion

The pyrolytic conversion of *A. indica* was performed in a tubular reactor at a temperature of 600 °C and pressure of 1 bar using nitrogen as inert conditions. The reactions were performed at a heating rate of 10 °C/min, and the pyrolytic conversion of *A. indica* indicated a yield of 25.574, 40.481 and 33.945 wt% of bio-oil, biochar and non-condensable gases, respectively. The organic phase had a calorific value of 19.62 MJ/kg with a density, viscosity and pH value of 1.053 g/ml, 0.933 mPa.s and 3.91, respectively. FTIR analysis of the organic phase indicated aliphatic compounds, ketones, ethers, alkyl compounds, aromatic compounds, hydrocarbons and alcohols. The biochar produced composed of fixed carbon (70.193 wt%) as its major constituent, and it had a calorific value 28.03 MJ/kg.

References

1. ClientEarth Communications, Fossil fuels and climate change: the facts What are fossil fuels ? What is the impact of fossil fuels on our planet ? How big is the impact of fossil fuels on climate change and our planet ? Can we continue to burn fossil fuel, (2020) pp. 5–9. https://www.clientearth.org/latest/latest-updates/stories/fossil-fuels-and-climate-change-the-facts/. Accessed 5 April 2022
2. Renewable Energy, Types, Forms & Sources, Types of renewable energy What, (n.d.). https://www.edfenergy.com/for-home/energywise/renewable-energy-sources. Accessed 5 April 2022
3. A. Demirbaş, Biomass resource facilities and biomass conversion processing for fuels and chemicals. Energy Convers. Manag. **42**, 1357–1378 (2001). https://doi.org/10.1016/S0196-8904(00)00137-0
4. L. Zhang, C. Xu, P. Champagne, Overview of recent advances in thermo-chemical conversion of biomass. Energy Convers. Manag. **51**, 969–982 (2010). https://doi.org/10.1016/j.enconman.2009.11.038
5. H.D. Kawale, N. Kishore, Pyrolysis of delonix regia and characterization of its pyrolytic products: effect of pyrolysis temperature. J. Energy Resour. Technol. **142**, 1–11 (2020). https://doi.org/10.1115/1.4046226
6. G. Ahmed, S. Acharya, H. Kawale, A. Singh, N. Kishore, S. Pal, Thermochemical conversion of Polyalthia longifolia leaves at different temperatures and characterization of their products. Fuel **280** (2020). https://doi.org/10.1016/j.fuel.2020.118574
7. A. Ahmed, M.S. Abu Bakar, R.S. Sukri, M. Hussain, A. Farooq, S. Moogi, Y.-K. Park, Sawdust pyrolysis from the furniture industry in an auger pyrolysis reactor system for biochar and bio-oil production. Energy Convers. Manag. **226,** 113502 (2020). https://doi.org/10.1016/j.enconman.2020.113502
8. H.D. Kawale, N. Kishore, Comparative study on pyrolysis of Delonix Regia, Pinewood sawdust and their co-feed for plausible bio-fuels production. Energy **203**, 117921 (2020). https://doi.org/10.1016/j.energy.2020.117921
9. Y. Sun, L. Liu, Q. Wang, X. Yang, X. Tu, Pyrolysis products from industrial waste biomass based on a neural network model. J. Anal. Appl. Pyrolysis. **120**, 94–102 (2016). https://doi.org/10.1016/j.jaap.2016.04.013
10. P. Das, M. Dinda, N. Gosai, S. Maiti, High energy density bio-oil via slow pyrolysis of Jatropha curcas shells. Energy Fuels **29**, 4311–4320 (2015). https://doi.org/10.1021/acs.energyfuels.5b00160

Kinetics and Thermodynamic Studies on Pyrolysis Behavior of *Peltophorum Pterocarpum* Via Thermogravimetric Analysis

Narra Thejaswini, Draksharapu Rammohan, and Nanda Kishore

Abstract The non-catalytic pyrolysis of *Peltophorum Pterocarpum* (PP) was carried out under an inert nitrogen gas atmosphere for a temperature limit of 25–900 °C by varying the heating rates between 10 and 55 °C/ min in the thermogravimetric analyzer. Model-free method, Kissinger–Akahira–Sunose (KAS) was utilized to calculate the kinetic parameters. The apparent activation energy, E_α (kJ/mol), and frequency factor, k_0 (min^{-1}), values of PP pyrolysis were estimated. Thermodynamic property such as change in enthalpy (ΔH, kJ/mol) was calculated from this model, and the average value obtained at 20 °C min^{-1} is 103. The reaction mechanism of PP during the pyrolysis process was analyzed by Criado's approach on the basis of KAS model. According to master plot results, the experimental results revealed the multistep reaction mechanisms of second order (R2), area and volume contracting (R2 and R3), and three-dimensional diffusion (D3) models at 10 °C min^{-1}.

Keywords *Peltophorum Pterocarpum* · Kissinger–Akahira–Sunose method · Thermogravimetric analysis · Kinetics · Thermodynamic properties

1 Introduction

In recent times, the idea of global warming has made people all over the world worry about how much energy they use and how well they use it. Even though energy efficiency is getting better all the time, the need for energy is growing faster. Fossil fuels-related emissions are a major danger to the environment of the world. Even though a lot of work is being done on a global scale, no one has yet found a way to control climate change. Renewable energy is laying the way for a planet that can last for a long time. Biofuel, solar, wind, and hydro are slowly taking the place of fossil fuels [1–4].

N. Thejaswini (✉) · D. Rammohan · N. Kishore
Department of Chemical Engineering, Indian Institute of Technology Guwahati, Guwahati, Assam, India
e-mail: n.thejaswini@iitg.ac.in

V. S. Moholkar et al. (eds.), *Sustainable Energy Generation and Storage*,
https://doi.org/10.1007/978-981-99-2088-4_13

Biofuels are produced from biodegradable, non-fossilized organic matter that comes from plants or animals. Often biomass is referred as lignocellulosic material because they contain lignin (12–30%), cellulose (31–55%), hemicellulose (20–35%), and other organic extractives [5]. Each of these materials has different thermal and physical properties [6–8]. Since biomass is less energy dense, burning it to get energy is not the best way to do it [9]. Thermochemical conversion can considerably boost energy density through pyrolysis and gasification processes. Pyrolysis is a thermochemical process that involves heating a substance to a high temperature in an anaerobic condition. Char, bio-oil, gas, and tar make up the product stream [10].

Pyrolysis begins with the determination of biomass properties, including apparent activation energy (E_α), frequency factor (k_0), and reaction model. Scaling up and designing pyrolysis reactors requires an understanding of the pyrolysis parameters [11, 12]. Computing E_α from biomass pyrolysis using current state-of-the-art methodologies requires first performing a thermogravimetric analysis (TGA) on the sample biomass and then performing computations utilizing model-free or model-fitting (iso-conversional) approaches [13, 14]. Model-fitting strategies have been shown to be less effective than iso-conversional (model-free) methods. Conducting TGA measurements under several non-isothermal situations has been established as the benchmark for iso-conversional kinetic. It is therefore feasible to accurately quantify E_α, k_0, and the reaction model by the use of iso-conversional methods. Thermal analysis is required in addition to kinetic analysis in order to determine the viability of a conversion process and to offer useful parameters for energy consumption [15, 16]. The kinetics and thermodynamic properties of PP biomass are determined in this study.

2 Samples and Methods

2.1 Sample Collection

Peltophorum Pterocarpum plant residues were collected from IIT Guwahati campus. Collected biomass was sundried. The dried biomass was then mechanically crushed with a cutting machine and sieved in a 60-mesh sieve to get the targeted particle size. The sieved biomass was stored in a plastic container for the further use of thermogravimetric operations.

2.2 Methods

In order to measure and record the mass loss of the sample during the pyrolysis reactions, the thermogravimetric device model TG 209 F1 (Netsch, Germany) was employed. A platinum crucible was inserted with approximately 6.8 ± 0.1 mg of

Peltophorum Pterocarpum powder. The sample was heated to 1073 K under four different dynamic heating regimes (10, 20, 35, and 55 °C/min) to determine the mass loss characteristics. A flow of 40 mL/ min of nitrogen was utilized for the entire pyrolysis process.

3 Theory

3.1 Kinetic Analysis

It is important to understand how temperature and conversion affect the kinetics of pyrolysis of biomass. Equation (1) is a typical way to express the overall conversion rate in this way:

$$\frac{d\alpha}{dt} = k(T)f(\alpha) \tag{1}$$

Here, α indicates degree of conversion, t stands for time of reaction, $k(T)$ denotes the rate constant as a function of temperature, and $f(\alpha)$ is the differential reaction model.

The dependency of rate constant $k(T)$ on temperature reported by Arrhenius is written as

$$k(T) = k_0 e^{\frac{-E\alpha}{RT}} \tag{2}$$

Here, k_0 is the frequency factor (min^{-1}), $E_\propto$ is the apparent activation energy at particular conversion (kJ/mol), and R is the universal gas constant.

Substituting Eq. (2) in Eq. (1):

$$\frac{d\alpha}{dt} = k_0 e^{\frac{-E\alpha}{RT}} f(\alpha) \tag{3}$$

Writing Eq. (3) in terms of heating rate (β):

$$\frac{d\alpha}{f(\alpha)} = \frac{k_0}{\beta} e^{\frac{-E\alpha}{RT}} dT \tag{4}$$

Integration of Eq. (4) results in

$$g(\propto) = \int_0^{\propto} \frac{d\alpha}{f(\alpha)} = \int_{T_0}^{T} \frac{k_0}{\beta} e^{\frac{-E_\propto}{RT}} dT \tag{5}$$

Rewritting Eq. (5) in terms of integration variable:

$$g(\propto) = \frac{k_0}{\beta}\int_0^T e^{\frac{-E_\propto}{RT}} = \frac{k_0 E_\propto}{\beta R} P(u) \tag{6}$$

Equation (6) is utilized by any iso-conversional method for obtaining kinetic parameters of any non-isothermal TGA data.

Kissinger–Akahira–Sunose method (KAS). Kissinger–Akahira–Sunose [17] approach is used to derive kinetic parameters. By substituting $p(u) = u^2 e^{-u}$ in Eq. (6):

$$\ln\left(\frac{\beta}{T^2}\right) = \ln\left(\frac{k_0 R}{E_\propto g(\propto)}\right) - \frac{E_\propto}{RT} \tag{7}$$

From Eq. (7), plot between $\ln\left(\frac{\beta}{T^2}\right)$ versus $\frac{1}{T}$ at four heating rates for the same conversion value will give slope and intercept. From the slope, i.e., $\frac{-E_\propto}{R}$, activation energy is obtained; from intercept, frequency factor is calculated.

3.2 *Thermodynamic Property*

Change in enthalpy is calculated with the obtained values of E_α from KAS method, frequency factor, and TGA data [13, 18].

$$\Delta H = E_\alpha \quad RT_m \tag{8}$$

3.3 *Criado's Master Plots: Reaction Mechanism*

Criado's master plot methodology was implemented to figure out reaction mechanism of given biomass. This Criado's method is categorized into three groups: differential ($f(\alpha)$), integral ($g(\alpha)$), and both ($z(\alpha)$) [19]. By multiplying the integral and differential of the reaction model, the $z(\alpha)$ master plot is obtained.

$$Z(\alpha)_{\text{experimental}} = g(\alpha)_{\text{experimental}} \times f(\alpha)_{\text{experimental}} \tag{9}$$

By Senum and Yang approximation, $p(u)$ is defined as

$$p(u) = \frac{e^{-u}}{u(u+4)\left(u^2+6u+6\right)} \tag{10}$$

Here, $u = \frac{E_\alpha}{RT}$

$$g(\alpha)_{\text{experimental}} = p(u) = \frac{k_0 E_\alpha}{\beta R}\left(\frac{e^{-u}}{u(u+4)\left(u^2+6u+6\right)}\right) \tag{11}$$

$$f(\alpha)_{\text{experimental}} = \left(\frac{d\alpha}{dt}\right)\left(\exp\left(\frac{E_\alpha}{RT}\right)\right) \tag{12}$$

$$Z(\alpha)_{\text{experimental}} = \left(\frac{d\alpha}{dt}\right)\left(\exp\left(\frac{E_\alpha}{RT}\right)\right) \times \frac{k_0 E_\alpha}{\beta R}\left(\frac{e^{-u}}{u(u+4)\left(u^2+6u+6\right)}\right) \tag{13}$$

$\alpha = 0.5$ is taken as reference; $z(0.5)$ can be represented as

$$Z(0.5)_{\text{experimental}} = \left(\frac{d0.5}{dt}\right)\left(\exp\left(\frac{E_{0.5}}{RT}\right)\right) \times \frac{k_0 E_{0.5}}{\beta R}\left(\frac{e^{-u}}{u(u+4)\left(u^2+6u+6\right)}\right) \tag{14}$$

Dividing Eq. (13) by Eq. (14), we get:

$$\frac{Z(\alpha)_{\text{experimental}}}{Z(0.5)_{\text{experimental}}} = \frac{\frac{d\alpha}{dt}\left(\exp\left(\frac{E_\alpha}{RT}\right)\right) \times \frac{k_0 E_\alpha}{\beta R}\left(\frac{e^{-u}}{u(u+4)\left(u^2+6u+6\right)}\right)}{\left(\frac{d0.5}{dt}\right)\left(\exp\left(\frac{E_{0.5}}{RT}\right)\right) \times \frac{k_0 E_{0.5}}{\beta R}\left(\frac{e^{-u}}{u(u+4)\left(u^2+6u+6\right)}\right)} \tag{15}$$

$$Z(\alpha)_{\text{theoretical}} = g(\alpha)_{\text{theoretical}} \times f(\alpha)_{\text{theoretical}} \tag{16}$$

Theoretical values are obtained from Eq. (16).

4 Results and Discussions

4.1 TGA Analysis

Thermogravimetric analysis of PP biomass was analyzed from room temperature to 900 °C at four different heating rates of $\beta = 10$, 20, 35 and 55 °C/min under inert nitrogen gas atmosphere. Degradation of PP biomass is categorized into three zones as shown in Fig. 1. Zone-I: Dehydration zone (25–200 °C): in this zone, only moisture removal happened, and weight loss of 7% is observed Zone-II. Active pyrolysis zone (220–410 °C), the main events of pyrolysis occur in this zone only. Degradation of high molecular weight components such as hemicellulose and cellulose occurs in this zone. Mass loss of 53–59% is observed in this zone; this was due to the breakdown of cellulose and hemicellulose. Next was passive pyrolysis zone (450–800 °C); in this Zone-II, lignin decomposition and char formation occur. Mass loss of 33% was

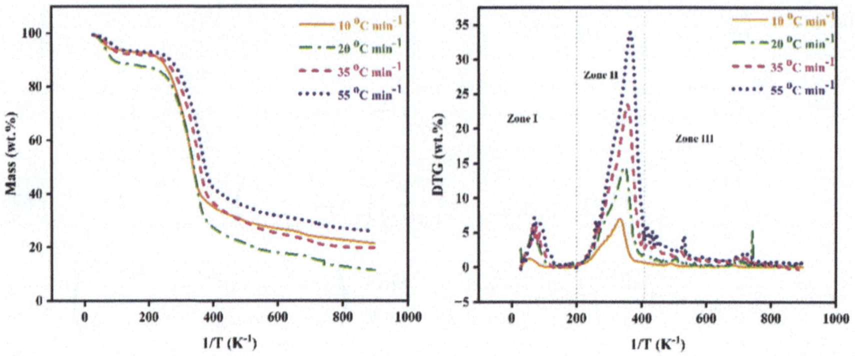

Fig. 1 Thermogravimetric and differential thermogravimetric profiles of PP biomass at different heating rates

observed in this zone. Similar thermogravimetric and differential thermogravimetric trends were observed for the other biomasses [2, 20, 21].

4.2 *Kinetic Analysis*

Calculation of activation energy and frequency factor KAS method was discussed in this section. Equation (7) was used to plot a graph between $\ln\left(\frac{\beta}{T^2}\right)$ versus $\frac{1}{T}$. Colored lines shown in Fig. 2 are called iso-conversional lines. Each line was drawn at four heating rates and single conversion. Each line gives slope and intercept; from the slope, activation energy was obtained, and intercept gives frequency factor At 0.1–0.8 conversions, eight activation energies were obtained and were listed in Table 1. The KAS approach yielded an average activation energy of 109.19 kJ/mol for *Peltophorum Pterocarpum* (PP) biomass. In this investigation, lower activation energy values (109.19 kJ/mol) were ascertained, then with other biomasses Delonix regia 205.67 kJ/mol [20], corn cob 240 kJ/mol [22], and poplar sawdust 159 kJ/mol [23]. The R^2 value of greater than 76% was obtained for all conversions showing that the thermal degradation data of PP biomass was properly fitted.

4.3 *Thermodynamic Analysis*

Activation energy obtained from KAS method was substituted in Eq. (8) to calculate the change in enthalpy (ΔH) values and given in Table 1. The average value of enthalpy at 10 °C min^{-1} was 104.65 kJ mol^{-1}. All the reported ΔH values were positive showing the endothermic nature of the biomass pyrolysis [20, 21]. From the analysis, it was noted that the trend for change in enthalpy and activation energy

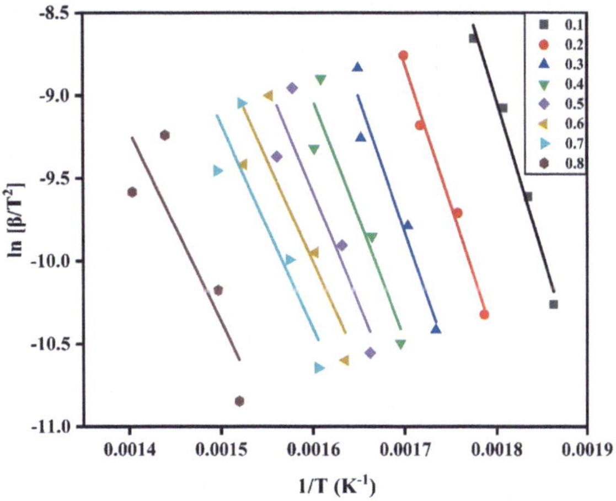

Fig. 2 Iso-conversional lines versus 1/T plot for KAS method

Table. 1 Calculated values of apparent activation energy and enthalpy at different conversions

Conversion (α)	Activation energy (E_α, kJ/mol)	Change in enthalpy (ΔH, kJ/mol)
0.1	84.32	79.79
0.2	144.78	140.25
0.3	133.65	129.11
0.4	118.18	113.65
0.5	109.94	105.41
0.6	99.48	94.94
0.7	97.98	93.45
0.8	85.20	80.66
Average	109.19	104.65

was in similar trend, and difference between these two properties was observed to be 4–5 kJ mol^{-1} indicating that there was no (or a very small, i.e., negligible) potential barrier stopping the creation of products [20, 21].

4.4 *Reaction Mechanism Prediction*

Criado's master plot methodology was applied for determining reaction mechanism. Using Eq. (15), experimental values were obtained and theoretical values were obtained from Eq. (16). According to literature, these theoretical models were categorized into five types [24]. Each type has different formulas for $f(\alpha)$ and $g(\alpha)$. Obtained experimental and theoretical values were plotted against conversion, as shown in Fig. 3. Whichever theoretical plot coincides with experimental curve at

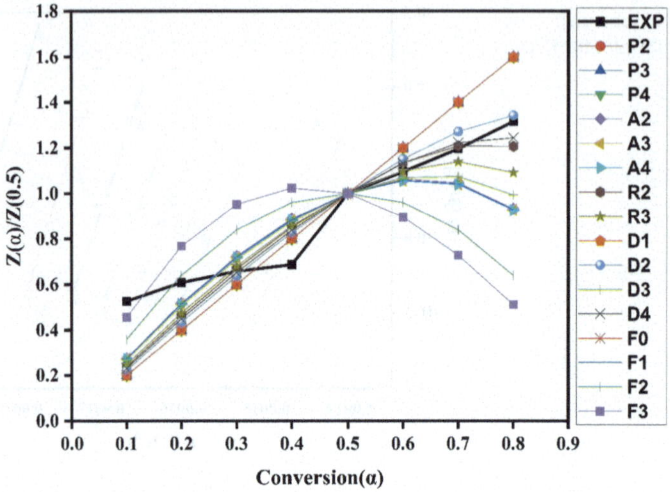

Fig. 3 Criado's master plot at 10 °C min^{-1}

corresponding conversion that reaction mechanism was occurring at that particular conversion. For PP biomass at 10 °C min^{-1} heating rate reaction till 0.8 conversion, the experimental curve followed several reaction mechanisms including second order (R2), area and volume contracting (R2 and R3), and three-dimensional diffusion (D3) models. From these findings, it was evident that PP biomass went through multistep reaction mechanism.

5 Conclusions

The major pyrolysis events occurred in the temperature range from 220 to 410 °C for PP biomass. Average activation energy obtained from KAS method was 108.75 kJ mol^{-1}. Variations in apparent activation energy were found to be caused by reactions that took place in multiple steps. Criado's master plot methodology revealed that PP biomass was in complex nature. Change in enthalpy calculations revealed the amount of external energy requirement for decomposition of PP biomass.

References

1. H.D. Kawale, N. Kishore, Bio-oil production from a lignocellulosic biomass and its fuel characteristics. J. Phys. Conf. Ser. **1276**(1) 2019. https://doi.org/10.1088/1742-6596/1276/1/012073
2. H.D. Kawale, N. Kishore, Comprehensive study on thermochemical putrefaction of *Delonix regia* in non-catalytic, catalytic and hydro-catalytic pyrolysis atmospheres. Renewable Energy

173, 223–236 (2021). https://doi.org/10.1016/j.renene.2021.03.139

3. V. Chintala, Production, upgradation and utilization of solar assisted pyrolysis fuels from biomass—a technical review. Renew. Sustain. Energy Rev. **90**(x), 120–130 (2018). https://doi.org/10.1016/j.rser.2018.03.066
4. A. Foley, A.G. Olabi, Renewable energy technology developments, trends and policy implications that can underpin the drive for global climate change. Renew. Sustain. Energy Rev. **68**, 1112–1114 (2017). https://doi.org/10.1016/j.rser.2016.12.065
5. Z. Anwar, M. Gulfraz, M. Irshad, Agro-industrial lignocellulosic biomass a key to unlock the future bio-energy: a brief review. J. Radiat. Res. Appl. Sci. **7**(2), 163–173 (2014). https://doi.org/10.1016/j.jrras.2014.02.003
6. W.H. Chen, K.M. Lu, W.J. Lee, S.H. Liu, T.C. Lin, Non-oxidative and oxidative torrefaction characterization and SEM observations of fibrous and ligneous biomass. Appl. Energy **114**, 104–113 (2014). https://doi.org/10.1016/j.apenergy.2013.09.045
7. N. Ottman et al., Soil exposure modifies the gut microbiota and supports immune tolerance in a mouse model. J. Allergy Clin. Immunol. **143**(3), 1198-1206.e12 (2019). https://doi.org/10.1016/j.jaci.2018.06.024
8. S. Wang, G. Dai, H. Yang, Z. Luo, Lignocellulosic biomass pyrolysis mechanism: a state-of-the-art review. Prog. Energy Combust. Sci. **62**, 33–86 (2017). https://doi.org/10.1016/j.pecs.2017.05.004
9. H. Weldekidan, V. Strezov, G. Town, T. Kan, Production and analysis of fuels and chemicals obtained from rice husk pyrolysis with concentrated solar radiation. Fuel **233**(June), 396–403 (2018). https://doi.org/10.1016/j.fuel.2018.06.061
10. H.D. Kawale, N. Kishore, Comparative study on pyrolysis of *Delonix regia*, Pinewood sawdust and their co-feed for plausible bio-fuels production. Energy **203**, 117921 (2020). https://doi.org/10.1016/j.energy.2020.117921
11. J. Chen, Y. Wang, X. Lang, X. Ren, S. Fan, Evaluation of agricultural residues pyrolysis under non-isothermal conditions: thermal behaviors, kinetics, and thermodynamics. Biores. Technol. **241**, 340–348 (2017). https://doi.org/10.1016/j.biortech.2017.05.036
12. G.K. Gupta, M.K. Mondal, Kinetics and thermodynamic analysis of maize cob pyrolysis for its bioenergy potential using thermogravimetric analyzer. J. Therm. Anal. Calorim. **137**(4), 1431–1441 (2019). https://doi.org/10.1007/s10973-019-08053-7
13. D. Rammohan, N. Kishore, R.V.S. Uppaluri, Reaction kinetics and thermodynamic analysis of non-isothermal co-pyrolysis of *Delonix regia* and tube waste. Bioresour. Technol. Rep. **18**(January), 101032 (2022). https://doi.org/10.1016/j.biteb.2022.101032
14. Z. Yao, S. Yu, W. Su, W. Wu, J. Tang, W. Qi, Kinetic studies on the pyrolysis of plastic waste using a combination of model-fitting and model-free methods. Waste Manage. Res. **38**(1_suppl), 77–85 (2020). https://doi.org/10.1177/0734242X19897814
15. J.L.F. Alves, J.C.G. da Silva, G.D. Mumbach, R.F. de Sena, R.A.F. Machado, C. Marangoni, Prospection of catole coconut (*Syagrus cearensis*) as a new bioenergy feedstock: insights from physicochemical characterization, pyrolysis kinetics, and thermodynamics parameters. Renewable Energy **181**, 207–218 (2022). https://doi.org/10.1016/j.renene.2021.09.053
16. S.L. Badshah, Z. Shah, J.L. Francisco Alves, J.C. Gomes da Silva, A. Iqbal, Pyrolysis of the freshwater macroalgae *Spirogyra crassa*: evaluating its bioenergy potential using kinetic triplet and thermodynamic parameters. Renewable Energy **179**, 1169–1178 (2021). https://doi.org/10.1016/j.renene.2021.07.105
17. H.E. Kissinger, Reaction kinetics in differential thermal analysis. Anal. Chem. **29**(11), 1702–1706 (1957). https://doi.org/10.1021/ac60131a045
18. A.K. Varma, S. Singh, A.K. Rathore, L.S. Thakur, R. Shankar, P. Mondal, Investigation of kinetic and thermodynamic parameters for pyrolysis of peanut shell using thermogravimetric analysis. Biomass Convers. Biorefinery (2020). https://doi.org/10.1007/s13399-020-00972-y
19. J.M. Criado, J. Málek, A. Ortega, Applicability of the master plots in kinetic analysis of non-isothermal data. Thermochim. Acta **147**(2), 377–385 (1989). https://doi.org/10.1016/0040-6031(89)85192-5

20. D. Rammohan, N. Kishore, R.V.S. Uppaluri, Insights on kinetic triplets and thermodynamic analysis of *delonix regia* biomass pyrolysis. Bioresour. Technol. **358**(May), 127375 (2022). https://doi.org/10.1016/j.biortech.2022.127375
21. R.K. Singh, D. Pandey, T. Patil, A.N. Sawarkar, Pyrolysis of banana leaves biomass: physico-chemical characterization, thermal decomposition behavior, kinetic and thermodynamic analyses. Bioresour. Technol. **310**(April), 123464 (2020). https://doi.org/10.1016/j.biortech.2020.123464
22. S. Singh, T. Patil, S.P. Tekade, M.B. Gawande, A.N. Sawarkar, Studies on individual pyrolysis and co-pyrolysis of corn cob and polyethylene: thermal degradation behavior, possible synergism, kinetics, and thermodynamic analysis. Sci. Total Environ. **783**, 147004 (2021). https://doi.org/10.1016/j.scitotenv.2021.147004
23. Y. Li, D. Yellezuome, R. Liu, J. Cai, Y. Gao, Investigation of product selectivity and kinetics of poplar sawdust catalytic pyrolysis over bi-metallic Iron-Nickel/ZSM-5 catalyst. Bioresour. Technol. **349**(January), 126838 (2022). https://doi.org/10.1016/j.biortech.2022.126838
24. P.E. Sánchez-Jiménez, L.A. Pérez-Maqueda, A. Perejón, J.M. Criado, Generalized master plots as a straightforward approach for determining the kinetic model: the case of cellulose pyrolysis. Thermochim. Acta **552**, 54–59 (2013)

Review on Thermal Management System of Li-Ion Battery for Electric Vehicle

Puneet Kumar Nema, P. Muthukumar, and Ranjith Thangavel

Abstract With the advent of electric vehicles (EVs) in the modern era to attain a sustainable environment, carbon neutrality, and energy-efficient solutions, demand for Li-ion batteries increased drastically. Lithium-ion batteries are preferred over other batteries for application in electric vehicles owing to their high energy density and long cycle life. However, the safety concerns due to thermal runaway issues, excessive heat generation during the fast charge/discharge process, and cycle life degeneration at high temperatures are severe setbacks for EVs commercialization. The cycle life and the performance of the Li-ion battery pack greatly depend on the temperature of operation, and so the operating temperature of the battery packs must be at the optimum range. The battery thermal management system (BTMS) plays a crucial role in maintaining the temperature of the battery packs at an optimum level to achieve the best performance and high safety. This review article provides a comprehensive summary of the recent developments in several battery thermal management systems (BTMS). This includes air-based, liquid-based, heat pipe, phase-change material, and the combination of these cooling methods. The advantages and disadvantages of the several BTMS are also compared and discussed. This review article can provide new insights for future energy researchers for designing an efficient BTMS for next-generation EVs.

Keywords Li-ion battery · Battery thermal management · Electric vehicle · Heat pipe · Battery safety · Phase-change material

P. K. Nema · P. Muthukumar · R. Thangavel (✉)
School of Energy Science and Engineering, Indian Institute of Technology Guwahati, Guwahati, Assam 781039, India
e-mail: ranjith.t@iitg.ac.in

P. Muthukumar
Department of Mechanical Engineering, Indian Institute of Technology Guwahati, Guwahati, Assam 781039, India

V. S. Moholkar et al. (eds.), *Sustainable Energy Generation and Storage*,
https://doi.org/10.1007/978-981-99-2088-4_14

1 Introduction

According to the World Energy Outlook report, since the industrial revolution, the average global surface temperature of the earth has increased by 1.1 °C due to a consistent increase in CO_2 level. In the early 1800s, the level of CO_2 was consistently less than 280 ppm as compared to 421 ppm in May 2022. To limit global warming and surface temperature within 1.5 °C, as specified in the Paris Agreement, emission levels must be slashed by 45% by 2030 and achieve net zero by 2050 [1]. To avoid the possible impact of climate change and achieve net zero, the world must immediately reduce carbon emissions, caused by the use of fossil fuels for transportation and power production, deforestation, and agriculture.

With the rapid growth in internal combustion engines, burning of fossil fuels leads to increased concentration of CO_2, together with other air pollutants such as NOx, smog, SOx, and particulates. In this context, several nations all over the world have initiated cooperative programs to tackle the consequences of climate change brought on by dangerous carbon dioxide emissions and greenhouse gases [2]. In a way to reduce emissions and mitigate health hazards, numerous studies on environment-friendly and sustainable energy sources have been conducted [3]. Based on prior research, electric vehicles (EVs) have been extensively perceived as a promising replacement for internal combustion engine vehicles, which most often create significant air pollution. EVs are clean and environment-friendly vehicles with nearly zero greenhouse gas emissions, more energy-efficient, and require less maintenance. The focus of the transportation industry is currently shifting toward electric vehicles (EVs). As per Global Electric Vehicle Outlook 2022 report, electric vehicle (EV) sales are nearly 6.6 million in 2021, nearly twice from the previous year. Only 120,000 electric vehicles were sold globally in 2012, while in 2021, these many are sold weekly. The total number of electric vehicles is approximately 16.5 million; that is, signifying the focus is changing toward EVs, which would be the key source of future transit [4].

The rising popularity and adoption of EVs are important for minimizing pollution, addressing the oil crisis, and achieving long-term economic and societal development. Hybrid electric vehicles (HEV), plug-in hybrid electric vehicles (PHEV), and pure electric vehicles (EVs) are the main types of EVs that are currently popularized [5]. For any electric vehicle, power sources, electric motor, transmission system, and electronic control system are primary components. The evolution of EVs largely depends on the development of energy-storage devices used as a power source. One of the most promising means of storing energy is the use of batteries. Recently, innovative battery technologies with great advantages including a lower self-discharge rate and a better energy density have evolved, offering promising energy-storage methods [6]. Due to higher volumetric energy density (400 Wh/L), higher specific energy density (150 Wh/kg), and longer lifespan, rechargeable lithium-ion (Li-ion) batteries are regarded as the best energy-storage option for electric vehicles as compared to other available batteries such as lead-acid, nickel–cadmium (Ni–Cd), and nickel-metal hydride (Ni–MH) [7].

The lithium-ion battery has positioned itself as one of the industry norms in the domain of electric vehicles. The battery functions by creating a potential difference between the positive electrode and negative electrode, submerged in an electrolyte, which is a conducting ionic liquid. It has a negative electrode made of carbon and a positive electrode made of lithium-containing compounds such as $LiFePO_4$, $LiMn_2O_4$, $LiCoO_2$ and $Li(Ni_xCo_yMn_z)O_2$. The lithium-ions are released from the negative electrode and travel through the electrolyte to reach the positive electrode during the discharging process, and the charging procedure follows opposite operation. These processes involve an electrochemical reaction and are followed by energy and heat generation [8]. The heat generated inside the battery during the charging and discharging process leads to temperature rise.

The performance parameters of lithium-ion batteries are dependent on the temperature at which the cells are working [6]. The functionality and lifespan of batteries might be negatively impacted by temperature changes above or below the range that is advised. Additionally, temperature differences among modules in a pack can have an adverse effect on the electrical balance, which lowers the performance of the battery pack. Since the battery has a greater specific energy, it produces an enormous amount of heat, which causes the lithium-ion cell to enter in an uncontrollable, self-heating state known as thermal runaway [9, 10]. Thermal runaway happens in batteries due to higher temperatures caused by heat-generating exothermic processes or could be due to short circuit. A significant increase in the temperature can potentially cause other destructive reactions. As a result, if heat is not adequately dispersed, the internal temperature of the battery rises rapidly. For Li-ion battery storage and operation, the upper temperature should be typically below 60 °C. Around 70–90 °C, the cell starts degrading, and it reaches near self-heating temperature leading to degradation of SEI (solid-electrolyte interface) layer between anode and electrolyte [11]. Temperature above 120 °C could fail the separator used between anode and cathode to prevent direct contact, and at temperature > 150 °C, cathode material reacts with electrolyte, producing oxygen that further reacts with electrolyte solvents. During thermal runaway, temperature of the cell could go beyond 600 °C, leading to fire and explosion [12].

All the reactions and associated heat generation leading to thermal runaway and battery explosion could be prevented by maximizing heat dissipation, usage of safer electrolytes, safer electrode materials, and proper battery thermal management technique (BTMS), considering the safety of Li-ion batteries. The main function of a thermal management system is to manage even temperature in the battery and to keep it within the appropriate operating range. An effective BTMS is critical for an EV battery pack to achieve optimal performance and have a long lifespan. It is crucial to minimize heat accumulations and remove the possibility of overheating. For the majority of conventional batteries, the ideal operating cell temperature varies between 15 °C and 40 °C which is nearby room temperature [13].

Numerous studies on battery thermal management (BTMS) have been conducted to propose technologies and methodologies for controlling the temperature range of battery cells, hence enhancing their functionality and efficiency [10, 14–17]. A thermal management system is required to standardize the batteries to work in the correct temperature range and minimize uneven temperature variation. For the battery to operate safely and effectively over an extended period, the ideal BTMS would be able to maintain the battery temperature within a reasonable range and achieve uniform temperature. For Li-ion batteries, safety must be examined at all levels, including the cell, module, pack, as well as vehicle. This is because failure at one level will lead to significantly more severe failures at a higher level [18]. To determine their suitability for prospective implementation in a modular system, various thermal management strategies that can be used to control the thermal behavior of Li-ion battery packs are assessed in this review article.

2 Battery Thermal Management System

An essential and fundamental component of a battery management system is the BTMS. Its primary purpose is to maintain the temperature of the battery cells in a pack within a desirable range. It contributes to extending the battery pack's lifespan while maintaining its secure and safe operation. The BTMS must be designed to meet automotive requirements, which include being lightweight, conveniently packaged in the desired application (e.g., EVs), compact, dependable, cost-effective, easy to assemble, and placed in an appropriate environment.

To address the Li-ion battery's need for heat dissipation, numerous cooling solutions in the active, passive, and hybrid forms are investigated. To manage the heat generated by batteries while they are in use, thermal management systems (TMS) are employed with air and liquid cooling systems, heat pipes, and phase-change material (PCM). Figure 1 represents the various BTMS methods used for cooling Li-ion batteries. Arora et al. [9] reviewed research work on the selection of optimal thermal management system based on their effectiveness, performance, and safety of EV. Author compared various available techniques for battery thermal management and found that a hybrid cooling system is more effective than compared to the individual cooling method. Based on ease of integration and energy efficiency, PCM would provide better cooling in a modular battery pack. However, PCM-based cooling methods have certain drawbacks of having additional weight and low thermal conductivity.

Choudhari et al. [15] reviewed the mechanism and effect of heat generation (Fig. 2) on the capacity of Li-ion battery, cathode, anode, electrolyte, and separator and further explained the reversible and irreversible heat generation phenomenon. Reversible heat is entropic heat generated at cathode and anode during electrochemical reaction, while irreversible heat has a contribution of above 70% at higher charging and discharging rate, characterized by joule heating and depends on C-rate. The author also signified the effect of airflow configuration and battery pack structure in air-cooling, the number of channels in liquid cooling, and the addition of graphene, CNT, and metal foam in PCM cooling to enhance thermal conductivity.

Zhao et al. [19] presented the various hybrid BTMS to meet the cooling requirements of electric vehicles and compared with various basic methods to highlight

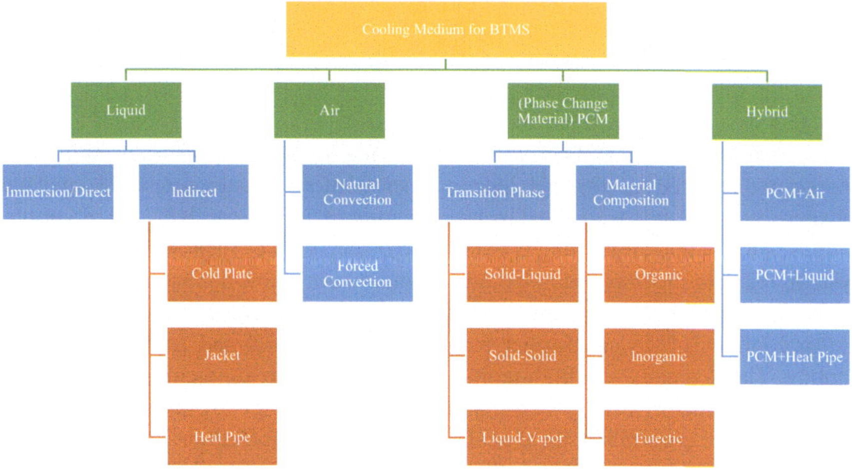

Fig. 1 Classification of various battery thermal management systems

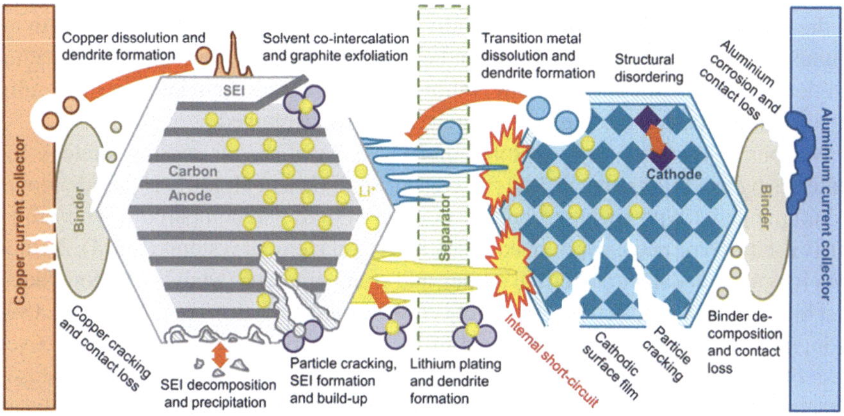

Fig. 2 Effect of heat generation in lithium-ion battery [15], reproduced with permission

its importance and better performance considering cost and efficiency. For hybrid cooling, the author considered heat pipe and PCM coupled with air or liquid cooling, respectively, PCM coupled with heat pipe, liquid cooling with air cooling, and thermoelectric cooling with basic BTMS. For extreme working conditions, the hybrid cooling system is more dependent and provides better cooling, but at the same time, this includes higher cost and makes the system more compact. Thus, a compromise should be made between cost and performance for better thermal management of EVs.

2.1 Air Cooling

One of the main cooling methods used to increase the efficiency and safety of EVs and HEVs is air cooling. It has the benefit of being simpler to pack, easier to maintain, having no risk of extra cooling fluid entering the battery pack, consuming less energy, and reduced weight [20]. Due to its straightforward structure and somewhat inexpensive price, it is well-liked in several commercial EV and HEV BTMS applications. An efficient air-cooling BTMS could disperse excess heat within the battery system, regulate the maximum operation temperature below a prescribed level, and maintain the maximum temperature variations within the necessary range [21].

Air-cooling BTMS can further be classified as natural and forced convection cooling. The demands of a high-temperature working environment, greater battery pack cooling, and high charge–discharge cycles were not met by natural cooling. To overcome these constraints, there is also the addition of forced-air cooling methods such as fans, blowers, and fins. Lithium-ion battery system for HEVs with forced-air cooling was also numerically studied using the thermal resistance model [22]. Figure 3 represents the schematic of the passive and active air-cooling system. Yang et al. [23] developed a thermal model to investigate the cell temperature variation in the battery pack. They computationally analyzed the 10S6P battery pack with an aligned and staggered configuration and experimentally validated the model with a single cell. For the longitudinal interval of 34 mm and transverse interval of 32 mm, maximum forced-air cooling of the battery pack was obtained.

Various researchers have explored and investigated the air-cooling strategy for batteries by modifying the airflow patterns [25–30]. Liu et al. [30] proposed a novel technique and J-type air-based battery cooling system and compared it with previously used U-type and Z-type air-based thermal management systems (Fig. 4). Author develops a battery electro-thermal model, and after parametric analysis, he concluded that his novel technique could effectively cool the battery pack and can obtain 31.18% reduction in temperature as compared to U- and Z-type system with the effectiveness of 35.3% and 46.6%, respectively. The suggested cooling technique increases temperature uniformity, decreases temperature fluctuation among cells, and reduces the total air flow requirement. Zhao et al. [24] reviewed the basic air-cooled BTMS and explained its benefits due to its low cost, simple design, and high reliability. He further compared active and passive air cooling and investigated the most recent advancements in EV and HEV air-cooling technologies by introducing the innovative design of air cooling and novel concepts of the battery pack. The author reviewed various in-line cell arrangements, ring-shaped, rectangular, and square battery packs. Among all, ring type gives the worst performance, and the square type provides optimum performance in terms of cost and cooling effectiveness. The addition of fins could also improve cooling performance, but basic air cooling could not effectively cool the battery pack at higher C-rate charge–discharge, high battery temperature, and thermal runaway conditions due to lower thermal conductivity. To enhance the performance under higher temperature conditions, air cooling could be coupled with

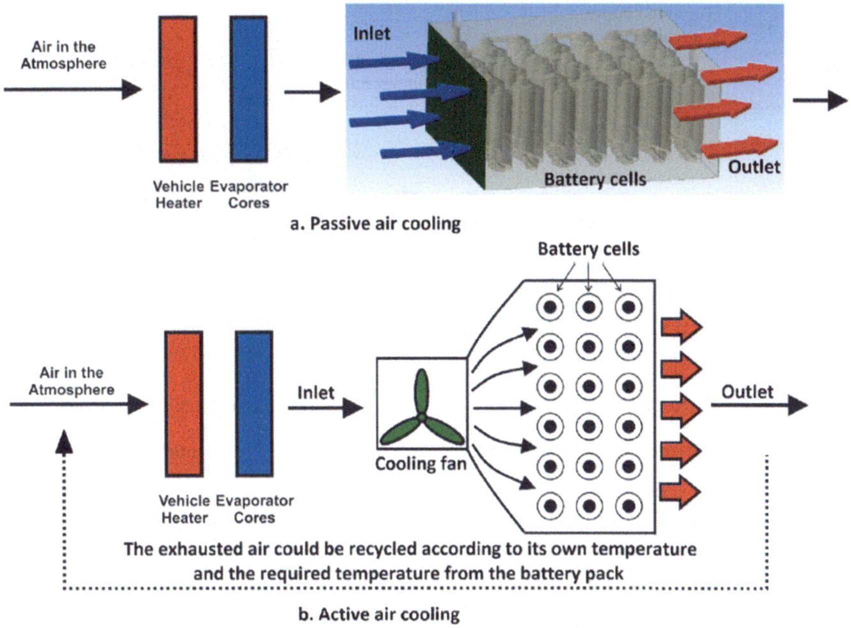

Fig. 3 Passive and active air-cooling technique [24], reproduced with permission

liquid cooling, PCM-based cooling, or some innovative design optimizations could be made.

2.2 *Liquid Cooling*

For high discharging rates, the cooling capacity of air-cooled BTMS seldom met the requirements. As compared to air cooling, liquid cooling can produce a higher cooling effect, better heat transfer, and can save almost 40% energy. The process of liquid cooling involves a coolant like water, mineral oil, refrigerant, dielectric fluid, or ethylene glycol to cool the batteries. It either travels through tubes, cold plates, and channels that are placed around the cells, or the cell could be immersed in the coolant to transfer excessive heat generated from the battery. The liquid cooling method can be classified as direct and indirect types of cooling as shown in Figure 5 [31].

Indirect cooling system

An indirect cooling system is considered one of the most effective cooling methods for EV BTMS owing to no direct contact between coolant and battery cell, uniform heat removal, and good thermal stability. Water and ethylene glycol mixture are typically used as a coolant that flows around the cell through tubing or channels. Various EV manufacturers such as Tesla, Chevrolet, Audi, and Volvo use indirect

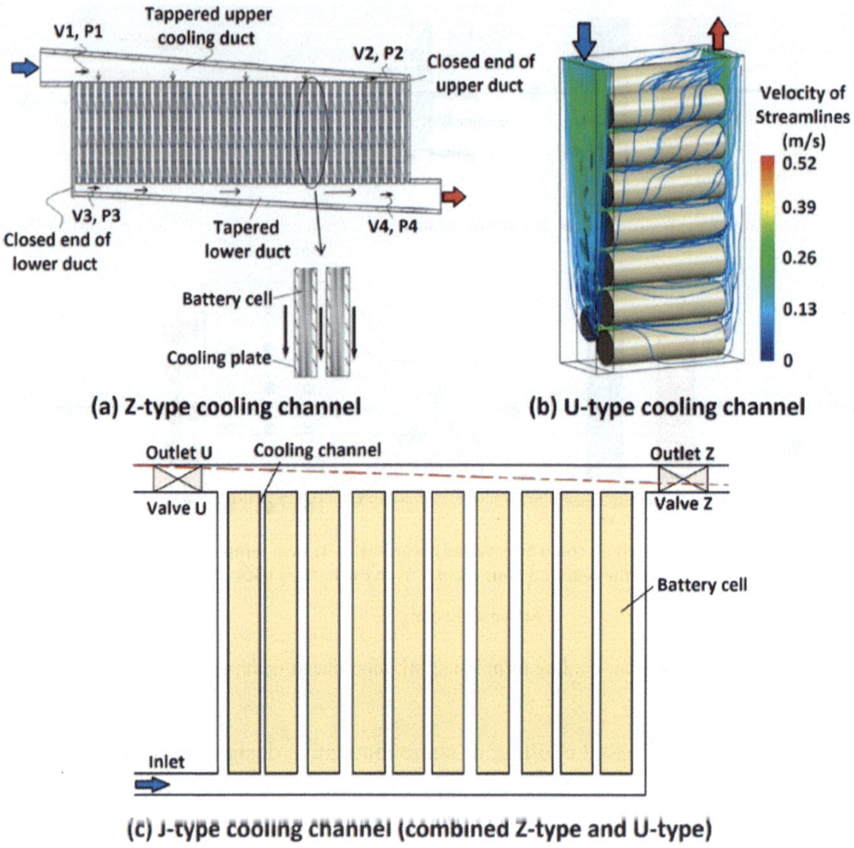

Fig. 4 Different cooling channel designs [24], reproduced with permission

liquid cooling techniques for their battery pack cooling as it provides better safety performance.

Research on indirect liquid cooling involves the investigation of coolants with higher thermal performance and the design of liquid coolant channels. Deng et al. [32] reviewed the performance of coolants like water, oil, liquid metal, and nanofluids and estimated their effectiveness based on different cooling strategies. They further investigated the use of nanoparticles to enhance the thermal conductivity of base coolants by comparing parameters such as the size of nanoparticles and their volume fraction. To enhance the heat transfer rate from the battery, shape of the channel, the position of the cooling plate, flow path geometry, and number of channels play an important role. The flow of coolant could be either serpentine type, U-turn type, or multichannel type. The movement of coolant through serpentine and multichannel types provides a larger heat transfer area as compared to U-type.

Though the indirect cooling system is considered an effective method for battery thermal management, certain limitations need to be addressed. Owing to the flow

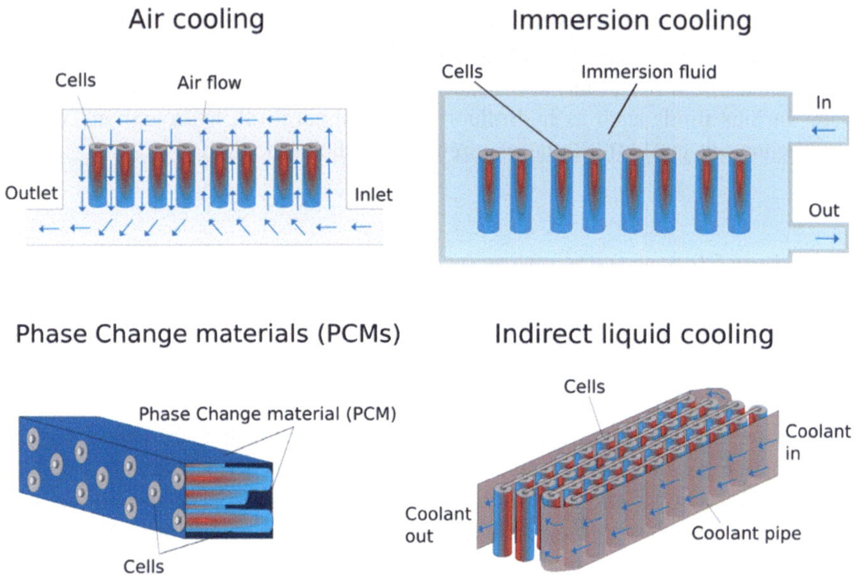

Fig. 5 Various thermal management systems [31], reproduced with permission

of liquid through a separate channel, the system becomes compact, extra weight is added, and leakage problems can persist. Apart from this, the presence of extra tubing between coolant and cell increases thermal resistance and could hamper heat removal capacity.

Direct cooling system/immersion cooling

The direct cooling method has grabbed attention recently for cooling lithium-ion battery packs in the EV industry. In direct or immersion cooling, the battery pack is immersed in dielectric fluid making direct contact so that the pack can have a uniform temperature and can efficiently remove heat at the cell level. This method has potential advantages over indirect cooling in terms of the more homogeneous path for heat removal as it reduces the thermal resistance between cell and liquid. The dielectric fluid used in immersion cooling can also suppress thermal runaway as it has flame retardant properties making the operation safer and more reliable [31].

Recent research on immersion cooling is mainly focused on the selection of dielectric fluid and performance analysis of battery packs under the immersed liquid. Patil et al. [33] experimentally and computationally investigated the cooling performance of lithium-ion battery pack immersed in a dielectric fluid and compared the cooling performance with other cooling methods. Results showed a 46.8% reduction in temperature compared to air cooling and a 9.3% drop in maximum temperature compared to indirect cooling for a 50 V battery pack at a 5 C discharge rate. Author also performed experiments for thermal abuse conditions and observed that immersed cooling could also reduce the risk of thermal runaway. For safer operation under immersion cooling, dielectric fluid with high thermal conductivity, higher specific heat, high fire point, non-volatile, non-toxic, and low viscosity should be used. Various fluids such as hydrofluoroethers, mineral oil, hydrocarbons, esters, and silicone oils could fulfill the requirements of dielectric fluid [31].

2.3 Heat Pipe

The concept of heat pipe was introduced by Richard. S. Gaugler in 1942 and was published in 1944 as a patent [34]. Extensive research work on heat pipes was first started by Los Alamos National Laboratory, New Mexico. Employing phase-change heat transfer, a heat pipe is a device that can run on its own without the need for externally supplied power and move enormous amounts of heat energy over long distances at rapid rates. This is true even for small temperature differences. Additionally, it offers attributes such as a compact structure, flexible shape, a long lifespan, and little maintenance, making it useful for effective heat control in numerous sectors. During the heat transfer process, initially, heat is conducted to the heat pipe in the evaporation section. Heat is then transferred to the condensation zone using a vapor line where condensation of working fluid is done. Since heat pipe has a flexible geometry and long life, it is used in EV battery cooling. Figure 6 presents the fundamental working cycle of heat pipe [35].

Ye et al. [36] did experiments on battery pack with a constant current of 18A and compared the performance of battery pack with and without heat pipe. Results showed that with the application of micro heat pipe arrays, the temperature differences in a cell were within 5 °C. Zhao et al. [37] developed a thermal management system employing a heat pipe with a wet cooling system. Experiment results showed that the proposed BTMS can work with stability under unsteady discharging conditions.

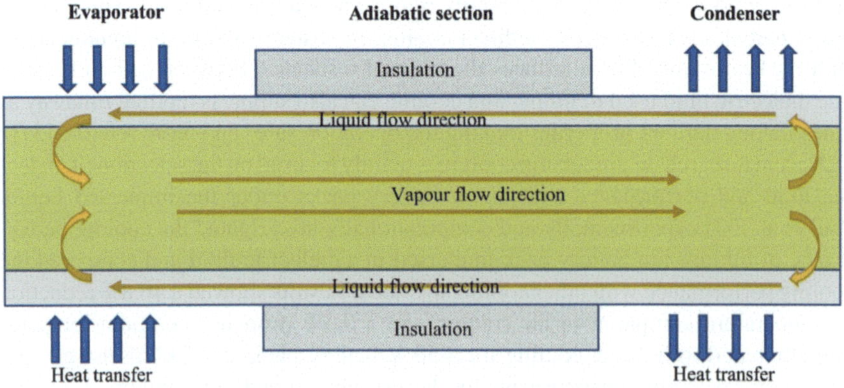

Fig. 6 Working principle of heat pipe [35], reproduced with permission

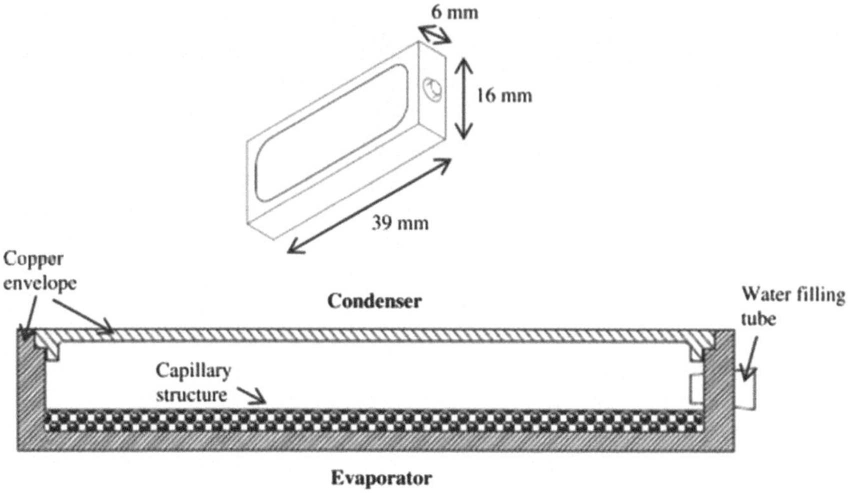

Fig. 7 Schematic of heat pipe experimental setup [39], reproduced with permission

Zhang et al. [38] developed a 3-D numerical model with a flat heat pipe BTMS and proposed a novel combination of heat pipe with fins for better thermal management. Tran et al. [39] experimentally justified the use of flat heat pipe for effective battery cooling as it reduces the thermal resistance by 30% and 20% as compared to air cooling and low-velocity cooling, respectively. Figure 7 shows the experimental setup used by them. Recently Abdelkareem et al. [40] focused on research work done on heat pipes and presented novel configurations of heat pipes with nanofluids and phase-change materials.

2.4 Phase-Change Material

Phase-change material is a material that can absorb or release latent heat to maintain a nearly constant temperature and is considered to be an attractive possibility for Li-ion battery temperature management. The application of PCM for BTMS was first proposed by Hallaj et al. [41] in a numerical study on a Sony US18650 cell with 100 Ah capacity using the finite element method. Paraffin wax was used as PCM that acts as a heat sink for heat generated during discharge, and after thermal simulation of the cylindrical cell, a uniform temperature was obtained providing better cooling of battery pack than that without PCM.

Based on transition type, PCM can be classified as solid–liquid, solid–vapor, and liquid–vapor. Solid–liquid-type PCM is generally used in EV battery cooling because of low volume change during the phase-change process [42]. Based on material composition, PCM can be further classified as organic, inorganic, and eutectic. Organic PCM such as paraffin has the advantage of chemical durability, stability, and corrosion resistance, but at the same time has low thermal conductivity. Inorganic PCM includes hydrated salts, melted salts, and metal and has high latent heat capacity. Eutectic PCMs are generally formed by blending two or more low melting temperature substances, such as organics, inorganics, or both inorganic and organic compounds, having a high melting point and high cost [43].

The PCM-based cooling technology has recently seemed to be the major choice for BTMS due to its quality of no extra equipment, simple operation, and relatively inexpensive. The large magnitude of latent heat allows PCM to store and dissipate energy, allowing the battery pack to remain within a safe operating temperature range for a longer period. Hemery et al. [44] experimentally investigated that PCM-based BTMS could effectively maintain temperature uniformity without failure as compared to natural and force air cooling in a 27-cell battery module of type 26,650. During the melting of commercial paraffin wax (Rubitherm RT28 HC), it can keep the temperature of the battery below 60 degrees Celsius. Ramandi et al. [45] computationally evaluated that the double-layer PCM had a greater exergy efficiency than the single-layer PCM. Using second law analysis, author estimated 50% exergy efficiency for single-layer PCM and 30–40% for two layers. The area of contact between both the battery and the PCMs had an impact on the system's thermal performance.

However, due to the low thermal conductivity and low heat transfer coefficient, heat transfer during a phase change is lowered, and the effectiveness of the PCM-based BTMS is reduced. To address the issue of low thermal conductivity, researchers are investigating strategies to improve the cooling capacity of battery packs by incorporating PCM with high thermal conductivity material or employing a PCM-based hybrid system. Various researchers used graphite, graphene, carbon nano-fiber, graphene nano-platelets, and multi-walled carbon nanotubes to enhance the thermal conductivity of PCM for more effective battery cooling [42]. Lazrak et al. [49] suggested a novel method to enhance the thermal conductivity of PCM by using copper grids and could reduce the temperature of the pack by a minimum of 5 °C. Author initially integrated different PCM with varying thermal conductivity in battery pack using 3-D configuration and further did an experimental study on a small-scale model. They found that PCM with the highest thermal conductivity was able to remove more heat from the cell to the surrounding. Lei et al. [50] proposed a novel hybrid BTMS integrating phase-change material, heat pipe, and spray cooling and experimentally demonstrated improvement in the thermal state of Li-ion battery for a wider range of temperature variations. This hybrid system consists of hydrated salt as PCM, a sintered water-copper pipe as a heat pipe attached with fins in the condenser end, and a spray loop kept in an acrylic box. The temperature of the battery pack varied from −40 °C to 150 °C. Results showed that this BTMS design can control battery temperature and maintain uniformity during high-temperature operation, but at the same time, could increase the weight and complexity of the system.

The existing thermal management techniques are not sufficient to meet the heat dissipation requirements of high energy density lithium ion batteries. In upcoming years hybrid BTMS combining two or more cooling methods will serve the purpose and solve the thermal runaway problems of the battery pack. Table 1 shows the comparative review, merits and demerits of cooling methods used for battery thermal management system.

3 Conclusion

Battery thermal management systems play an important role in the development of modern-era lithium-ion batteries as it improves the safety of vehicle, performance, and life of the battery. In this review article, various battery thermal management techniques such as air, liquid, heat pipe, and PCM are discussed with their advantages and operating principles. The air-cooling method is preferred because of its low cost, simpler and safer operation, but for higher discharge rate, air cooling with various airflow patterns is found to be the most optimized method to improve air-cooling BTMS. Performance enhancement can also be done by designing and using novel airflow patterns and by the utilization of fins to improve convective heat transfer. As compared to air cooling, direct and indirect method of liquid cooling can produce a higher cooling effect and better heat transfer due to the higher thermal conductivity of liquid as compared to air. In recent years, leading EV manufacturers are adopting indirect-type liquid cooling with varying channel geometry such as serpentine and mini channel flow that can provide a better cooling effect. Further research is required on coolants in direct cooling method as it is showing enormous potential in terms of homogenous heat removal from a battery pack.

The phase-change material based cooling method is capable of removing a huge quantity of heat due to its latent heat capacity with no external energy consumption. However, pure PCM has limitations to remove a large quantity of heat generated inside the battery due to its low thermal conductivity. The addition of high thermal conductivity materials such as graphene, CNT, metal foam, and nanomaterials to PCMs may further improve the heat transfer rate. PCM-based hybrid cooling techniques combining passive cooling like PCM with active cooling such as air or liquid cooling and PCM with heat pipe have high potential to be developed as next-generation cooling methods and focus research in this area is required. The use of heat pipe for BTM is comparably new for EV, and its combination with other cooling methods or nanofluids can be investigated for a better and more effective battery pack.

Table 1 Comparison of various battery thermal management techniques

S. No.	Cooling method	Battery geometry/configuration	Nominal voltage (V)/capacity (Ah)	Max. temp. (°C)	ΔT_{max} (°C)	Merits	Demerits	Author	Remarks
1	Natural air cooling	10S6P, cylindrical 18,650 lithium-ion cells	37 V/15.6 Ah	51.7	–	• Low initial cost • No operative cost • Easy to integrate • Less weight	• Low heat transfer coefficient • Limited temperature reduction • Uneven temperature distribution • Low efficiency • Not suitable for high C-rates • Low efficiency • Not suitable for high C-rates	Jiaqiang et al. [21]	Efficiency could be increased by changing the flow pattern

(continued)

Table 1 (continued)

S. No.	Cooling method	Battery geometry/configuration	Nominal voltage (V)/capacity (Ah)	Max. temp. (°C)	ΔT_{max} (°C)	Merits	Demerits	Author	Remarks
2	Forced-air cooling	48 prismatic lithium-ion battery cells	55 Ah	19.17 W thermal power at 2C		• Simple operation • Low initial cost • Low maintenance	• Extra energy consumption • Low heat transfer coefficient • Leakage problem	Xu et al. [27]	U-type airflow pattern
3	Indirect liquid cooling	4 cylindrical 18,650 cells	–	40	6.78	• Better thermal conductivity than air • Good cooling capacity	• Proper designing of the liquid flow channel and a cooling jacket are required	Yates et al. [46]	Micro-/mini channels for proper cooling are used
4	Direct/immersion liquid cooling	14 lithium-ion pouch cells	50 V	< 40	46.8% temperature drop	• Direct fluid contact with all cell surfaces • Compact structure • Higher cooling capacity	• Low heat capacity of dielectric fluid • Material complexity • Leakage problem	Patil et al. [33]	A highly efficient technique for high-capacity battery cooling and currently in the R&D stage

(continued)

Table 1 (continued)

S. No.	Cooling method	Battery geometry/configuration	Nominal voltage (V)/capacity (Ah)	Max. temp. (°C)	ΔT_{max} (°C)	Merits	Demerits	Author	Remarks
5	Heat pipe	15 lithium-ion prismatic cells	34.5 V/23 Ah	38.2	–	• Compact system, low weight • High heat dissipation efficiency • Low cost and low maintenance	• Temperature inhomo-geneity in the battery pack • High initial cost	Behi et al. [47]	Thermal performance is investigated at different discharge rates
		5 prismatic lithium-ion cells	16 V	38	< 5			Zhang et al. [38]	
6	Pure PCM	27 cylindrical 26,650 lithium-ion cells	–	< 60	–	• Low energy consumption • Affordable cost • Uniform temperature distribution	• Low thermal conductivity • Bulky system • Not effective for higher C-rates	Hemery et al. [44]	PCM is effective as compared to natural air cooling, but adds weight and cost
7	Hybrid cooling (Heat pipe-assisted PCM)	5 prismatic batteries	–	42.8	–	• Uniform temperature • Better performance as compared to the pure PCM method	• Relatively longer operation time	Wu et al. [48]	Various combinations of cooling methods could be used

References

1. IEA, World Energy Outlook 2021: part of the World Energy Outlook. Int. Energy Agency 386 (2021) [Online]. Available: https://www.iea.org/reports/world-energy-outlook-2021
2. B.R. David, S. Spencer, J. Miller, S. Almahmoud, H. Jouhara, Comparative environmental life cycle assessment of conventional energy storage system and innovative thermal energy storage system. Int. J. Thermofluids **12**(September), 100116 (2021). https://doi.org/10.1016/j.ijft.2021.100116
3. P.A. Owusu, S. Asumadu-Sarkodie, A review of renewable energy sources, sustainability issues and climate change mitigation. Cogent Eng. **3**(1) (2016). https://doi.org/10.1080/23311916.2016.1167990
4. IEA, Global EV outlook 2022 securing supplies for an electric future, p. 221 (2022) [Online]. Available: https://iea.blob.core.windows.net/assets/ad8fb04c-4f75-42fc-973a-6e54c8a4449a/GlobalElectricVehicleOutlook2022.pdf
5. Q. Wang, B. Jiang, B. Li, Y. Yan, A critical review of thermal management models and solutions of lithium-ion batteries for the development of pure electric vehicles. Renew. Sustain. Energy Rev. **64**, 106–128 (2016). https://doi.org/10.1016/j.rser.2016.05.033
6. A.G. Olabi et al., *Battery Energy Storage Systems and SWOT (Strengths, Weakness, Opportunities, and Threats) Analysis of Batteries in Power Transmission*, vol. 254 (Elsevier Ltd, 2022). https://doi.org/10.1016/j.energy.2022.123987
7. S.M.A.S. Bukhari, J. Maqsood, M.Q. Baig, S. Ashraf, T.A. Khan, Comparison of characteristics-lead acid, nickel based, lead crystal and lithium based batteries. in *Proceeding—UKSim-AMSS 17th International Conference Computer Modelling Simulation, UKSim 2015*, pp. 444–450 (2016). https://doi.org/10.1109/UKSim.2015.69
8. X. Zhang, Z. Li, L. Luo, Y. Fan, Z. Du, A review on thermal management of lithium-ion batteries for electric vehicles. Energy **238**, 121652 (2022). https://doi.org/10.1016/j.energy.2021.121652
9. S. Arora, Selection of thermal management system for modular battery packs of electric vehicles: a review of existing and emerging technologies. J. Power Sources **400**(July), 621–640 (2018). https://doi.org/10.1016/j.jpowsour.2018.08.020
10. J. Kim, J. Oh, H. Lee, Review on battery thermal management system for electric vehicles. Appl. Therm. Eng. **149**(November 2018), 192–212 (2019). https://doi.org/10.1016/j.applthermaleng.2018.12.020
11. R. Spotnitz, J. Franklin, Abuse behavior of high-power, lithium-ion cells. J. Power Sources **113**(1), 81–100 (2003). https://doi.org/10.1016/S0378-7753(02)00488-3
12. C. Mikolajczak, M. Kahn, K. White, R.T. Long, Lithium-Ion battery failures 43–70 (2011). https://doi.org/10.1007/978-1-4614-3486-3_4
13. Q. L. Yue, C.X. He, M.C. Wu, T.S. Zhao, Advances in thermal management systems for next-generation power batteries. Int. J. Heat Mass Transf. **181** (2021). https://doi.org/10.1016/j.ijheatmasstransfer.2021.121853
14. L. Yang, F. Zhou, L. Sun, S. Wang, Thermal management of lithium-ion batteries with nanofluids and nano-phase change materials: a review. J. Power Sources **539**(March), 231605, (2022). https://doi.org/10.1016/j.jpowsour.2022.231605
15. V.G. Choudhari, D.A.S. Dhoble, T.M. Sathe, A review on effect of heat generation and various thermal management systems for lithium ion battery used for electric vehicle. J. Energy Storage **32**(March), 101729 (2020). https://doi.org/10.1016/j.est.2020.101729
16. A.G. Olabi et al., Battery thermal management systems: recent progress and challenges **15**(June) (2022). https://doi.org/10.1016/j.ijft.2022.100171
17. Universitat Politècnica de València, Battery Thermal Management Systems (BTMSs) (2019)
18. D. Doughty, E.P. Roth, A general discussion of Li Ion battery safety. Electrochem. Soc. Interface **21**(2), 37–44 (2012). https://doi.org/10.1149/2.F03122if
19. C. Zhao, B. Zhang, Y. Zheng, S. Huang, T. Yan, X. Liu, Hybrid battery thermal management system in electrical vehicles: a review. Energies **13**(23) (2020). https://doi.org/10.3390/en13236257

20. A.A. Pesaran, S. Burch, M. Keyser, An approach for designing thermal management systems for electric and hybrid vehicle battery packs preprint. Fourth Veh. Therm. Manag. Syst. Conf. Exhib. **24–27**(January), 1–18 (1999)
21. E. Jiaqiang et al., Effects of the different air cooling strategies on cooling performance of a lithium-ion battery module with baffle. Appl. Therm. Eng. **144**(May), 231–241 (2018). https://doi.org/10.1016/j.applthermaleng.2018.08.064
22. H. Park, A design of air flow configuration for cooling lithium ion battery in hybrid electric vehicles. J. Power Sources **239**, 30–36 (2013). https://doi.org/10.1016/j.jpowsour.2013.03.102
23. N. Yang, X. Zhang, G. Li, D. Hua, Assessment of the forced air-cooling performance for cylindrical lithium-ion battery packs: a comparative analysis between aligned and staggered cell arrangements. Appl. Therm. Eng. **80**, 55–65 (2015). https://doi.org/10.1016/j.applthermaleng.2015.01.049
24. G. Zhao, X. Wang, M. Negnevitsky, H. Zhang, A review of air-cooling battery thermal management systems for electric and hybrid electric vehicles. J. Power Sources **501**(April) (2021). https://doi.org/10.1016/j.jpowsour.2021.230001
25. T. Wang, K.J. Tseng, J. Zhao, Z. Wei, Thermal investigation of lithium-ion battery module with different cell arrangement structures and forced air-cooling strategies. Appl. Energy **134**, 229–238 (2014). https://doi.org/10.1016/j.apenergy.2014.08.013
26. K. Chen, S. Wang, M. Song, L. Chen, Configuration optimization of battery pack in parallel air-cooled battery thermal management system using an optimization strategy. Appl. Therm. Eng. **123**, 177–186 (2017). https://doi.org/10.1016/j.applthermaleng.2017.05.060
27. X.M. Xu, R. He, Research on the heat dissipation performance of battery pack based on forced air cooling. J. Power Sources **240**, 33–41 (2013). https://doi.org/10.1016/j.jpowsour.2013.03.004
28. K. Yu, X. Yang, Y. Cheng, C. Li, Thermal analysis and two-directional air flow thermal management for lithium-ion battery pack. J. Power Sources **270**, 193–200 (2014). https://doi.org/10.1016/j.jpowsour.2014.07.086
29. K. Chen, M. Song, W. Wei, S. Wang, Structure optimization of parallel air-cooled battery thermal management system with U-type flow for cooling efficiency improvement. Energy **145**, 603–613 (2018). https://doi.org/10.1016/j.energy.2017.12.110
30. Y. Liu, J. Zhang, Design a J-type air-based battery thermal management system through surrogate-based optimization. Appl. Energy **252**(March), 113426 (2019). https://doi.org/10.1016/j.apenergy.2019.113426
31. C. Roe et al., Immersion cooling for lithium-ion batteries—a review. J. Power Sources **525**(February), 231094 (2022). https://doi.org/10.1016/j.jpowsour.2022.231094
32. Y. Deng et al., Effects of different coolants and cooling strategies on the cooling performance of the power lithium ion battery system: a review. Appl. Therm. Eng. **142**(June), 10–29 (2018). https://doi.org/10.1016/j.applthermaleng.2018.06.043
33. M. Suresh Patil, J.H. Seo, M.Y. Lee, A novel dielectric fluid immersion cooling technology for Li-ion battery thermal management. Energy Convers. Manag. **229**(December 2020), 113715 (2021). https://doi.org/10.1016/j.enconman.2020.113715
34. R.S. Gaugler, 'Heat Transfer Device', U. S. Patent 2,350,348, p. 5 (1944) [Online]. Available: https://patentimages.storage.googleapis.com/2b/dd/b3/5f4d19cee50bf8/US2350348.pdf
35. M.A. Abdelkareem et al., Heat pipe-based waste heat recovery systems: background and applications. Therm. Sci. Eng. Prog. **29**(January), 101221 (2022). https://doi.org/10.1016/j.tsep.2022.101221
36. X. Ye, Y. Zhao, Z. Quan, Experimental study on heat dissipation for lithium-ion battery based on micro heat pipe array (MHPA). Appl. Therm. Eng. **130**, 74–82 (2018). https://doi.org/10.1016/j.applthermaleng.2017.10.141
37. R. Zhao, J. Gu, J. Liu, An experimental study of heat pipe thermal management system with wet cooling method for lithium ion batteries. J. Power Sources **273**, 1089–1097 (2015). https://doi.org/10.1016/j.jpowsour.2014.10.007
38. Z. Zhang, K. Wei, Experimental and numerical study of a passive thermal management system using flat heat pipes for lithium-ion batteries. Appl. Therm. Eng. **166**(November 2019), 114660 (2020). https://doi.org/10.1016/j.applthermaleng.2019.114660

39. T.H. Tran, S. Harmand, B. Desmet, S. Filangi, Experimental investigation on the feasibility of heat pipe cooling for HEV/EV lithium-ion battery. Appl. Therm. Eng. **63**(2), 551–558 (2014). https://doi.org/10.1016/j.applthermaleng.2013.11.048
40. M.A. Abdelkareem et al., Thermal management systems based on heat pipes for batteries in EVs/HEVs **51**(March) (2022). https://doi.org/10.1016/j.est.2022.104384
41. S. Al Hallaj, J.R. Selman, A novel thermal management system for electric vehicle batteries using phase-change material **147**(9), 3231–3236 (2000)
42. J. Chen et al., Effects of different phase change material thermal management strategies on the cooling performance of the power lithium ion batteries: a review. J. Power Sources **442**(October), 227228 (2019). https://doi.org/10.1016/j.jpowsour.2019.227228
43. Y.E. Milián, A. Gutiérrez, M. Grágeda, S. Ushak, A review on encapsulation techniques for inorganic phase change materials and the influence on their thermophysical properties. Renew. Sustain. Energy Rev. **73**(December 2016), 983–999 (2017). https://doi.org/10.1016/j.rser.2017.01.159
44. C.V. Hémery, F. Pra, J.F. Robin, P. Marty, Experimental performances of a battery thermal management system using a phase change material. J. Power Sources **270**, 349–358 (2014). https://doi.org/10.1016/j.jpowsour.2014.07.147
45. M.Y. Ramandi, I. Dincer, G.F. Naterer, Heat transfer and thermal management of electric vehicle batteries with phase change materials. Heat Mass Transf. und Stoffuebertragung **47**(7), 777–788 (2011). https://doi.org/10.1007/s00231-011-0766-z
46. M. Yates, M. Akrami, A.A. Javadi, Analysing the performance of liquid cooling designs in cylindrical lithium-ion batteries. J. Energy Storage 33(June 2019), 100913 (2021). https://doi.org/10.1016/j.est.2019.100913
47. H. Behi et al., Thermal management analysis using heat pipe in the high current discharging of lithium-ion battery in electric vehicles. J. Energy Storage **32**(September), 101893 (2020). https://doi.org/10.1016/j.est.2020.101893
48. W. Wu, X. Yang, G. Zhang, K. Chen, S. Wang, Experimental investigation on the thermal performance of heat pipe-assisted phase change material based battery thermal management system. Energy Convers. Manag. **138**, 486–492 (2017). https://doi.org/10.1016/j.enconman.2017.02.022
49. A. Lazrak, J.F. Fourmigué, J.F. Robin, An innovative practical battery thermal management system based on phase change materials: numerical and experimental investigations. Appl. Therm. Eng. **128**, 20–32 (2018). https://doi.org/10.1016/j.applthermaleng.2017.08.172
50. S. Lei, Y. Shi, G. Chen, A lithium-ion battery-thermal-management design based on phase-change-material thermal storage and spray cooling. Appl. Therm. Eng. **168**(December 2019), 114792 (2020). https://doi.org/10.1016/j.applthermaleng.2019.114792

Reaction Kinetics of Non-isothermal Pyrolysis of Tube Waste in Thermogravimetric Analyzer

Draksharapu Rammohan, Nanda Kishore, and R. V. S. Uppaluri

Abstract The current study concentrated on the non-isothermal pyrolysis of tube waste (TW), emphasizing the detailed thermal decomposition and kinetic results. All the experiments were performed at temperature (°C) between 25 and 1000 with a nitrogen flow rate (mL min^{-1}) of 40 and at multiple heating rates (°C min^{-1}) 5, 10, 20, 35, and 55. To estimate the reaction kinetics, the Differential Friedman (DFM) and Distributed Activation Energy (DAE) iso-conversional (model-fitting and model-free) methods were employed. The reaction kinetic results including average kinetic value activation energy E_α (kJ mol^{-1}) and frequency factor k_0 (min^{-1}) were ascertained to be 244.73 and 2.73 × 1024 by the DFM (model-fitting) method and 223.37 and 1.84 × 1024 by the DAE (model-free) method.

Keywords Tube waste · TGA · Non-isothermal · Pyrolysis · Thermal decomposition · Kinetics

1 Introduction

Butyl rubber tube is used in an extensive range of applications, including inner vehicle tubes, automobile components, and rubber stoppers. These applications generate a significant volume of waste disposed of in environmentally harmful methods, such as combustion and incineration [1]. These applications generate a significant volume of waste which is disposed off by environmentally harmful methods, such as burning and incineration.

Thus, a waste disposal mechanism that is environmentally benign is required, and pyrolysis is one of the most promising technologies [2, 3]. Numerous research papers on pyrolysis using various iso-conversional methods for different feedstocks have been published as far as concern to biomass work. However, the detailed kinetics for the tube waste (TW) feedstock is not yet reported. Hence, the current research report

D. Rammohan (✉) · N. Kishore · R. V. S. Uppaluri
Department of Chemical Engineering, Indian Institute of Technology Guwahati, Guwahati, Assam, India
e-mail: drammohan@iitg.ac.in

V. S. Moholkar et al. (eds.), *Sustainable Energy Generation and Storage*,
https://doi.org/10.1007/978-981-99-2088-4_15

aims to comprehend the thermal decomposition analysis, followed by a detailed investigation of the kinetic study. The thermal decomposition and kinetic evolution of various tube waste materials, including butadiene and styrene rubbers (BR and SBR), have been studied elsewhere [4, 5]. The pyrolysis of tube waste is conducted at five heating rates (°C min^{-1}) including 5, 10, 20, 30, and 55, respectively, under an inert condition of nitrogen at 40 mL min^{-1} flow rate and temperature varying from 25 to 1000 °C. The two different (model-free and model-fitting) iso-conversional methods, such as Differential Friedman (DFM) and Distributed activation (DAE), are employed in the present study.

2 Material and Methods

2.1 Material

The tube waste (TW) material was collected from backyard source (cycle repair shop) inside IIT Guwahati campus. The material was washed with water, sundried for 24 h, and cut into the (< 250 μm) particle size. Then the purified TW was further used for physiochemical and thermogravimetric analysis (TGA) characterization.

2.2 Physiochemical Characterization

The TW material was subjected to several ASTM standards procedures (ASTM: E872-82, E1756-08, E1755-01) to assess the proximate analysis including volatile, moisture, and ash contents. In addition, the ultimate and higher heating value (HHV) analyses of TW were determined using a using a CHNS analyzer and oxygen bomb calorimetry. Furthermore, the physicochemical results were presented in our previous works [2, 6].

2.3 TGA

Thermal decomposition of TW was analyzed in TGA unit (TG209F1). The TGA was operated in a dynamic mode with the 5 heating rates (°C min^{-1}) 5, 10, 20, 35, and 55 between 25 and 1000 temperature (°C) under inert condition of nitrogen 40 mL min^{-1}. About 6 mg of TW sample weight was used for each experimental run.

2.4 Reaction Kinetics

Reaction kinetics of TW including activation energy, E_α (kJ mol^{-1}), and frequency factor, k_0 (min^{-1}), was estimated in the current research work. Two different iso-conversional methods including model-fitting and model-free were employed to determine the reaction kinetics. The pyrolysis reaction behavior of TW was described as follows:

$$\text{Tube Waste} \xrightarrow[\substack{T\,=\,25\,-\,1000\,^\circ\text{C} \\ N_2=\,40\ \text{mL/min}}]{k(T)} \text{Volatiles} + \text{Residue} \tag{1}$$

The TW conversion rate was defined as

$$\frac{\mathrm{d}\alpha}{\mathrm{d}t} = k(T) f(\alpha) \tag{2}$$

where the conversion rate, reaction rate, and reaction method of the TW were represented by $\frac{\mathrm{d}\alpha}{\mathrm{d}t}$, $k(T)$, and $f(\alpha)$, respectively.

The TW fractional conversion was given as

$$\alpha = \frac{m_i - m_t}{m_i - m_f} \tag{3}$$

where the initial, at any time, and final time corresponding weights of the TW were represented by m_i, m_f, and m_t, respectively.

The TW reaction rate was represented by Arrhenius expression as

$$k(T) = k_0 \exp\left(-\frac{E_\alpha}{\mathrm{RT}}\right) \tag{4}$$

where the activation energy (kJ mol^{-1}), frequency factor (min^{-1}), universal gas constant (J mol^{-1} K^{-1}), and temperature (K) of the TW were represented by E_α, k_0, R, and T, respectively.

The following equation was obtained from above Eqs. 2 and 4:

$$\frac{\mathrm{d}\alpha}{\mathrm{d}t} = k_0 \exp\left(-\frac{E_\alpha}{\mathrm{RT}}\right) f(\alpha) \tag{5}$$

The heating rate term (β) was defined as

$$\beta = \frac{\mathrm{d}T}{\mathrm{d}t} = \frac{\mathrm{d}T}{\mathrm{d}\alpha} \times \frac{\mathrm{d}\alpha}{\mathrm{d}t} \tag{6}$$

From the above Eqs. 5 and 6, the resultant generalized expression ($g(\alpha)$) was

$$g(\alpha) = \int_0^{\alpha} \frac{d\alpha}{f(\alpha)} = \frac{k_0}{\beta} \int_{T_0}^{T} \exp\left(-\frac{E_\alpha}{RT}\right)$$
$$dT = \frac{k_0 E_\alpha}{\beta R} \int_x^{\infty} u^{-2} \exp^{-u} du = \frac{k_0 E_\alpha}{\beta R} p(x) \quad (7)$$

where the temperature approximation was represented by $p(x)$.

Equation 7 does not have the exact solution; however, it can be solved by utilizing the various iso-conversional methods (model-fitting and model-free) in order to obtain the E_α and k_0.

2.5 Model-Fitting Method: Differential Friedman (DFM)

The mathematical equation of DFM Friedman [7] is

$$\ln\left[\beta\left(\frac{d\alpha}{dT}\right)\right] = \ln\left[k_0 f(\alpha)^n\right] - \frac{E_\alpha}{RT_\alpha} \quad (\text{DFM} - 8)$$

A plot was drawn between $\ln\left[\beta\left(\frac{d\alpha}{dT}\right)\right]$ and $\frac{1}{T_\alpha}$, and the corresponding values of slope and intercept provide the E_α and k_0.

2.6 Model-Free Method: Distributed Activation Energy (DAE)

DAE Miura and Maki [8] method employed the temperature approximation $p(x) = 0.6075 - x$ and the mathematical equation of DAE as

$$\ln\left(\frac{\beta}{T_\alpha^2}\right) = \ln\left[\frac{k_0 R}{E_\alpha}\right] + 0.6075 - \left[\frac{E_\alpha}{RT_\alpha}\right] \quad (\text{DAE} - 9)$$

A plot was drawn between $\ln\left(\frac{\beta}{T_\alpha^2}\right)$ and $\frac{1}{T_\alpha}$, and the corresponding values of slope and intercept provide the E_α and k_0.

3 Results and Discussions

3.1 TGA

The TGA of tube waste (TW) was conducted at five heating rates (°C min^{-1}) including 5, 10, 20, 35, and 55 from 25 to 1000 °C temperature. Both TG and DTG profiles were shown in Fig. 1. From Fig. 1, it was observed that the decomposition of TW was categorized into three stages, namely stage 1, stage 2, and stage 3. In stage 1, maximum removal of water, plasticizers, and low molecular weight components were occurred, whereas in the stage 2, maximum decomposition of TW occurred with the weight loss of approximately 59 to 60 wt% between 5 and 55 °C min^{-1}. The highest rate of decomposition of TW was observed at temperatures of 359, 370, 381, 392, and 401 °C, respectively. The temperature change was caused by an increase in heating rate [9–12]. Furthermore, the highest rate of decomposition of TW was elevated at a higher heating rate (°C min^{-1}) of 50. The massive decomposition of TW, isobutylene (IB), and isoprene (IP) was represented by three peaks on DTG profiles for each heating rate. Similar TG and DTG profiles have been observed elsewhere [6, 13, 14]. In stage 3 at 5–55 °C min^{-1}, an undecomposed TW with a weight loss of 45–55% was noticed without any share peaks.

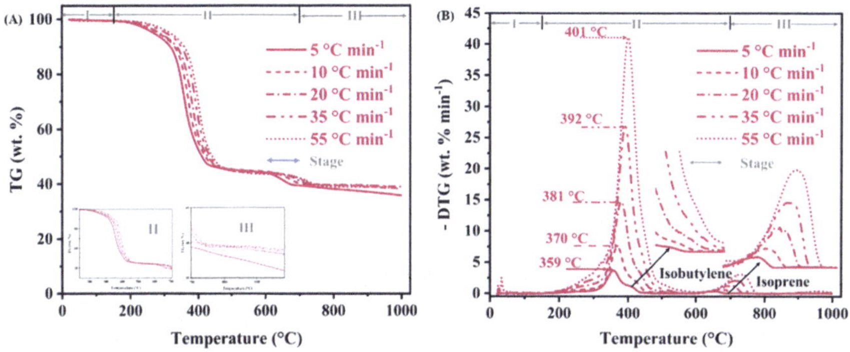

Fig. 1 Variation of activation energy by model-fitting (DFM) and model-free (DAE) methods for different conversion range. The following reactions were explained the mechanism during the pyrolysis process, and this stage was known as active pyrolysis

Reaction (1) TW $\rightarrow V1 + R1$ (Main chain splitting)

$$\text{Reaction (2)} \begin{cases} 2a) R1 \rightarrow V2 \\ 2b) R1 \rightarrow R2 \ \ (Depolymerization) \\ 2c) R2 \rightarrow V3 \end{cases}$$

Reaction (3) $R2 \rightarrow R3$ (Cyclization)

Reaction (4) $R3 \rightarrow V4 + R4$ (Degradation of cyclization products)

where the volatile matter of monomers, dimers, trimer, and tetra trimers of the TW polymeric components were represented by $V1$, $V2$, $V3$, and $V4$, respectively. The radical formation of H_2, CH_4; CO, CO_2; aliphatic, aromatic, H_2S; and solid char of the TW polymeric components were represented by $R1$; $R2$; $R3$; and $R4$, respectively.

Initially, at lower temperatures (< 250 °C), the TW main chain splits into $V1$ and $R1$ during reaction 1 [4, 15–17]. At intermediate temperatures (250–750 °C), reaction 2 continued the decomposition of reaction 1 [5, 15, 16, 18, 19]. Finally, at higher temperatures (750 °C), cyclization and cyclic degradation of reaction 2 to reaction 3 and reaction 4 occurred [5, 15, 16, 18, 19].

3.2 Reaction Kinetics

The reaction kinetics including activation energy (E_α) and frequency factor (k_0) of TW were ascertained from two different iso-conversional methods (model-fitting and model-free). The two methods DFM and DAE were used to estimate E_α and k_0 from 5 to 55 °C min^{-1}. The kinetics results were provided in Fig. 2 and Table 1. As depicted in Fig. 1, the kinetic values E_α (kJ mol^{-1}) and k_0 (min^{-1}) varied from 165.12 to 339.72 and from 1.56×10^{13} to 1.02×10^{25} for DFM; 143.47 to 332.94 and from 5.79×10^{12} to 1.48×10^{25} for DAE methods, respectively, between 0.1 and 0.8 conversion. The alterations of E_α and k_0 (10^{13} to 10^{16} and 10^{25}) increased between conversion 0.1 to 0.3 and 0.8 due to the decomposition of isobutylene and isoprene polymers by reaction 2 and reaction 3 during the pyrolysis of TW [15, 16]. From Table 1 and Fig. 2, the lowest average kinetic values E_α (kJ mol^{-1}) and k_0 (min^{-1}) were 223.37 and 1.84×10^{24}, respectively, obtained by the DAE (model-free) method. The highest average kinetic factors activation energies E_α (kJ mol^{-1}) and frequency factor k_0 (min^{-1}) were 244.73 and 2.73×10^{24}, respectively, ascertained by the DFM (model-fitting) method. A good correlation coefficient ($R^2 = 0.987$) was observed between the two methods; however, the DAE method yielded the best kinetic results. Using DFM method, the average kinetic values of butadiene and styrene rubbers (BR and SBR) were 200 and 240 kJ mol^{-1} published elsewhere [4, 5]. According to the kinetic reports, the BR has received 40 kJ than the SBR. The present best kinetics results of 223 kJ mol^{-1} were also compared with the other studies, and the results were consistent; it was also evident that the TW had a lower kinetic energy than the SBR from 0.1 to 0.8 conversion [5, 9, 20].

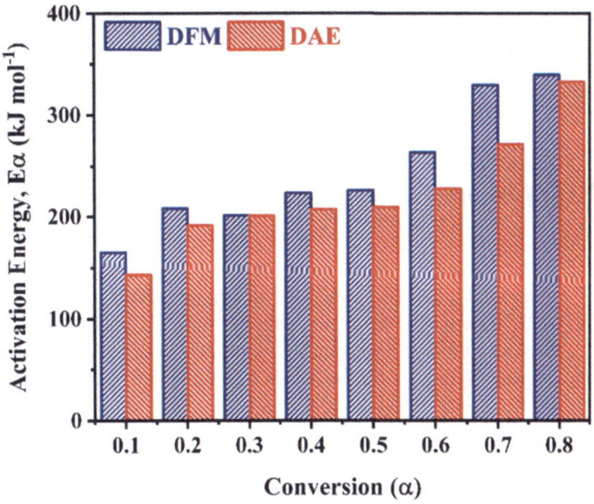

Fig. 2 Variation of activation energy by model-fitting (DFM) and model-free (DAE) methods for different conversion ranges

Table 1 Comparison of kinetic parameters obtained by model-fitting (DFM) and model-free (DAE) methods

α	Model-fitting (DFM) method			Model-free (DAE) method		
	E_α	k_0	R^2	E_α	k_0	R^2
0.1	165.12	1.56E+13	0.962	143.47	5.79E+12	0.970
0.2	208.12	3.06E+16	0.997	191.49	1.13E+16	0.989
0.3	201.54	7.56E+15	0.995	201.10	3.05E+16	0.994
0.4	223.90	4.32E+17	0.999	207.31	5.64E+16	0.998
0.5	226.26	4.83E+17	0.995	210.32	6.14E+16	0.998
0.6	263.59	2.51E+20	0.989	228.25	1.01E+18	0.994
0.7	329.59	1.16E+25	0.974	272.10	1.36E+21	0.983
0.8	339.72	1.02E+25	0.984	332.94	1.48E+25	0.969
Avg.	244.73	2.73E+24	0.987	223.37	1.84E+24	0.987

4 Conclusions

During the non-isothermal pyrolysis of TW, the maximum decomposition occurs at stage 2, with an approximately 59–60% weight loss between 5 and 55 °C min^{-1}. Increasing heating rates influence the peak decomposition temperature and decomposition rates. In comparison with the model-fitting method (DFM), the model-free method (DAE) demonstrated the most accurate kinetic results. The DAE (model-free) method yielded the lowest average kinetic values E_α (kJ mol^{-1}) and k_0 (min^{-1}) of 223.37 and 1.84×10^{24}, respectively. The variations in frequency factor from 5.79×10^{12} to 1.48×10^{25} for DAE method between 0.1 and 0.8 conversion indicated a faster decomposition rate of TW.

Acknowledgements Authors were thankful to North-East Research Conclave (NERC) 2022 and anonymous reviewers during the publication of the present research work.

References

1. EPA, *Advancing Sustainable Materials Management* (United Sates 2020). https://www.epa.gov/sites/default/files/2021-01/documents/2018_ff_fact_sheet_dec_2020_fnl_508.pdf
2. H.D. Kawale, N. Kishore, Thermochemical putrefaction of *Delonix regia* biomass and tube waste to produce high-quality pyrolytic bio-oil. J. Thermal Anal. Calorim. 0123456789 (2021). https://doi.org/10.1007/s10973-021-10725-2
3. O. Bičáková, P. Straka, Co-pyrolysis of waste tire/coal mixtures for smokeless fuel, maltenes and hydrogen-rich gas production. Energy Convers. Manage. **116**, 203–213 (2016). https://doi.org/10.1016/j.enconman.2016.02.069
4. M.G. Aboelkheir, L.Y. Visconte, G.E. Oliveira, R.D. Toledo Filho, F.G. Souza, The biodegradative effect of Tenebrio molitor Linnaeus larvae on vulcanized SBR and tire crumb. Sci. Total Environ. **649**, 1075–1082 (2019). https://doi.org/10.1016/j.scitotenv.2018.08.228
5. B. Danon, N.M. Mkhize, P. Van Der Gryp, J.F. Görgens, Combined model-free and model-based devolatilisation kinetics of tyre rubbers. Thermochim. Acta **601**, 45–53 (2015). https://doi.org/10.1016/j.tca.2014.12.003
6. D. Rammohan, N. Kishore, R.V.S. Uppaluri, Reaction kinetics and thermodynamic analysis of non-isothermal co-pyrolysis of *Delonix regia* and tube waste. Bioresour. Technol. Rep. **18**(January), 101032 (2022). https://doi.org/10.1016/j.biteb.2022.101032
7. H.L. Friedman, Kinetics of thermal degradation of char-forming plastics from thermogravimetry. Application to a phenolic plastic. J. Polym. Sci. Part C: Polym. Symposia **6**(1), 183–195 (1964). https://doi.org/10.1002/polc.5070060121
8. K. Miura, T. Maki, A simple method for estimating f(E) and k0(E) in the distributed activation energy model. Energy Fuels **12**, 864–869 (1998). https://doi.org/10.1021/ef970212q
9. J. Han, W. Li, D. Liu, L. Qin, W. Chen, F. Xing, Pyrolysis characteristic and mechanism of waste tyre: A thermogravimetry-mass spectrometry analysis. J. Anal. Appl. Pyrol. **129**(December 2017), 1–5 (2018). https://doi.org/10.1016/j.jaap.2017.12.016
10. R.M. Neves, H.L. Ornaghi, F.G. Ornaghi, S.C. Amico, A.J. Zattera, Degradation kinetics and lifetime prediction for polystyrene/nanocellulose nanocomposites. J. Therm. Anal. Calorim. **147**(1), 879–890 (2022). https://doi.org/10.1007/s10973-020-10316-7
11. T. Qiu, F. Ge, C. Li, S. Lu, Study of the thermal degradation of flame-retardant polyester GFRP using TGA and TG-FTIR-GC/MS. J. Therm. Anal. Calorim. **147**(10), 5743–5760 (2022). https://doi.org/10.1007/s10973-021-10895-z
12. A. Dziemidkiewicz, M. Maciejewska, M. Pingot, Thermal analysis of halogenated rubber cured with a new cross-linking system. J. Therm. Anal. Calorim. **138**(6), 4395–4405 (2019). https://doi.org/10.1007/s10973-019-08881-7
13. D. Pradhan, R.K. Singh, Characterization of the liquid product obtained by the pyrolysis of a bicycle tube. Energy Sources, Part A: Recovery, Utilization Environ. Eff. **37**(19), 2099–2106 (2015). https://doi.org/10.1080/15567036.2011.608106
14. R. Peñalver et al., Assessing the level of airborne polystyrene microplastics using thermogravimetry-mass spectrometry: results for an agricultural area. Sci. Total Environ. **787**, 147656 (2021). https://doi.org/10.1016/j.scitotenv.2021.147656
15. S. Kordoghli, B. Khiari, M. Paraschiv, F. Zagrouba, M. Tazerout, Impact of different catalysis supported by oyster shells on the pyrolysis of tyre wastes in a single and a double fixed bed reactor. Waste Manage. **67**, 288–297 (2017). https://doi.org/10.1016/j.wasman.2017.06.001
16. S. Kordoghli, M. Paraschiv, R. Kuncser, M. Tazerout, F. Zagrouba, Catalysts' influence on thermochemical decomposition of waste tires. Environ. Prog. Sustain. Energy **36**(5), 1560–1567 (2017). https://doi.org/10.1002/ep.12605

17. B. Qu et al., Kinetic analysis of waste tire pyrolysis with metal oxide and zeolitic catalysts. J. Anal. Appl. Pyrol. **152**(October) (2020). https://doi.org/10.1016/j.jaap.2020.104949
18. B. Khiari, S. Kordoghli, D. Mihoubi, F. Zagrouba, M. Tazerout, Modeling kinetics and transport phenomena during multi-stage tire wastes pyrolysis using Comsol®. Waste Manage. **78**, 337–345 (2018). https://doi.org/10.1016/j.wasman.2018.06.002
19. N. Nkosi, E. Muzenda, J. Gorimbo, M. Belaid, Developments in waste tyre thermochemical conversion processes: gasification, pyrolysis and liquefaction. RSC Adv. **11**(20), 11844–11871 (2021). https://doi.org/10.1039/d0ra08966d
20. F. Xu et al., TG FTIR and Py-GC/MS study on pyrolysis mechanism and products distribution of waste bicycle tire. Energy Convers. Manage. **175**(September), 288–297 (2018). https://doi.org/10.1016/j.enconman.2018.09.013

Kinetics and Thermodynamic Studies on Pyrolysis Behavior of Plastic Waste Using Thermogravimetric Analysis

Praveen Kumar Reddy Annapureddy, Draksharapu Rammohan, and Nanda Kishore

Abstract The purpose of this study was to investigate the thermochemical breakdown of polypropylene grocery bags by using non-isothermal thermogravimetric analysis. The temperature was varied from room temperature to 900 °C in a series of experiments at different heating rates. Differential Friedman method (DFM), an isoconversional model, is utilized in this study to evaluate the kinetic parameters for pyrolysis of polypropylene grocery bags. According to the DFM method, the obtained values of average apparent activation energy (E_α) and pre-exponential factor (k_0) were, respectively, 158.98 kJ mol^{-1} and 2.16×10^{12} min^{-1}. An $R^2 > 0.98$ correlation was found in the conversion range of 0.2–0.8 for DFM technique in the kinetics analysis. The DFM approach determined an average enthalpy change (ΔH, kJ/mol) of 153.12 kJ mol^{-1} at a heating rate of 10 K min^{-1}. Master plots based on the integral form of kinetic data were used to find out the best pyrolysis kinetic model for polypropylene grocery bags. The kinetic process of thermal dehydration of polypropylene was explained by the zero-order reaction mechanism (F0), which was based on the plots between experimentally and theoretically calculated master plots.

Keywords Polypropylene grocery bags · Differential Friedman method · Thermogravimetric instrument · Master plots

1 Introduction

Due to social and political concerns, large consumption of fossil fuels, and their environmental implications, it was projected that the overuse of petroleum, coal, and natural gas will be reduced in the near future. Contrarily, the amount of plastic trash generated by containers and packaging has increased to 60% in recent years, making it an increasingly important component. Approximately 70% of all plastic trash is

P. K. R. Annapureddy (✉) · D. Rammohan · N. Kishore
Department of Chemical Engineering, Indian Institute of Technology Guwahati, Guwahati, Assam, India
e-mail: annapureddy.pra@iitg.ac.in

V. S. Moholkar et al. (eds.), *Sustainable Energy Generation and Storage*,
https://doi.org/10.1007/978-981-99-2088-4_16

composed of polyolefins [1]. Only a small fraction of recycled material and most non-biodegradable plastics have contributed to a rise in trash output because of their widespread use. Technical solutions are needed to combat plastic waste's negative influence on the environment. Plastic garbage should not be disposed of in landfills or incinerated because both constitute a hazard to the environment by polluting the soil with leaching and soil impregnation materials, as well as creating pollutants, and neither alternative is ideal [2].

Kinetics has a big impact on how pyrolysis is designed for complex materials like plastic. A careful look at the values of apparent activation energy (E_α), the frequency factor (k_0), and reaction mechanism [3] is all crucial to pyrolysis kinetics precision. Thermal degradation properties and dynamics of plastic waste pyrolysis can be evaluated using thermogravimetric analysis because of its simplicity and high precision [4]. The main goal of kinetic analysis is to find out the kinetic parameters [5, 6]: apparent activation energy (E_α) and reaction mechanism and also to find the change in enthalpy (ΔH).

Isoconversional approaches, also called model-free approaches, are the best way to analyze TGA data because, unlike model-fitting methods, they don't assume a reaction mechanism. Differential methods and integral methods are the two types of isoconversional methods. The differential Friedman method (DFM), which is a differential method, was chosen for this work [7]. When performing energy calculations and establishing the viability of a process, it is critical to take thermodynamic properties into account as well [8]. The kinetic parameters derived by isoconversional methods can be used to calculate changes in enthalpy (ΔH). The kinetic models can be determined using master plot approaches at this point. For non-isothermal experiments, master plots based on integral form of the kinetic data were particularly useful because the activation energy value could be known [9].

The number of research addressing the kinetic triplet, thermodynamic calculations, and determining the reaction mechanism based on the integral form of the kinetic data together was found to be quite a few only in light of prior literature studies, and only a few studies employing different sources of polypropylene were observed. In this research work, authors explored the kinetic and thermodynamic studies for the pyrolysis of polypropylene grocery bags at different heating rates.

2 Materials and Methods

2.1 Materials

Polypropylene (PP) grocery bags were used in this study, and these bags were taken from local supermarket shop. The bags were then cut into small threads, crushed into powder with scissors, and sieved to get particles into a uniform size of 700 μm.

2.2 Methods

In a nitrogen-inert atmosphere, thermogravimetric analysis (TGA) was conducted on a PP sample in a non-isothermal environment. During the pyrolysis event, a TG 209 F1 (Netsch, Germany) thermogravimetric analyzer was used to monitor and track the sample's mass change. Approximately 6 mg of the sample was used in each experiment, and it was heated to 900 °C in the crucible at four different heating rates: 10, 20, 35, and 55 K min^{-1}.

3 Theory

3.1 Kinetic Analysis

Kinetic analysis is concerned with the modeling and evaluation of kinetic parameters using various isoconversional techniques. These approaches provide values for activation energy (E_α) that can be used to predict frequency factor (k_0) and to select a reaction mechanism with the help of master plots based on the integral form of the kinetic data. The general expression [9–12] for the kinetics of solid decomposition assumes the rate to be a function of only two factors, temperature and conversion, which were represented by the following equation:

$$\frac{d\alpha}{dt} = k(T)f(\alpha) \tag{1}$$

where α is the defined as conversion, expressed as the fraction of pyrolyzed polypropylene, T denotes the temperature, t denotes time, $k(T)$ is the rate constant which is a function of temperature, and it is given by Arrhenius equation: $k = k_0 e^{\frac{-E\alpha}{RT}}$ and $f(\alpha)$ indicates the model of reaction in differential form. Here k_0, E_α, and R represent apparent activation energy (kJ mol^{-1}), the frequency factor (min^{-1}) and gas constant, respectively.

α can be represented as:

$$\alpha = \frac{m_0 - m_t}{m_0 - m_f},$$

where m_0, m_t, and m_f denote the weight of the sample at the start, at time, t, and at the end of process. Equation (1) can be rearranged as follows by substituting the Arrhenius equation:

$$\frac{d\alpha}{dt} = k_0 e^{\frac{-E\alpha}{RT}} f(\alpha) \tag{2}$$

The temperature in non-isothermal TGA studies is dependent on time (t) and rises with a heating rate (β), $\beta = \frac{dT}{dt}$. Thus, by plugging $dt = \frac{dT}{\beta}$ into Eq. (2):

$$\frac{d\alpha}{f(\alpha)} = \frac{k_0}{\beta} e^{\frac{-E\alpha}{RT}} dT \tag{3}$$

The above Eq. (3) is the basic differential equation that can be utilized to derive kinetic parameters from non-isothermal TGA studies.

Differential Friedman Method

DFM [7] was a differential approach and one of the earliest examples of isoconversional methods, and this can be obtained by applying natural logarithm on the both terms of Eq. (3):

$$\ln\left(\beta \frac{d\alpha}{dT}\right) = \ln(k_0 g(\alpha)) - \frac{E\alpha}{RT} \tag{4}$$

Linear fitting approach was used for the DFM model by plotting of $\ln\left(\beta \frac{d\alpha}{dT}\right)$ versus $\frac{1}{T}$ graphs at varied heating rates with constant conversion values. The gradients of these straight lines $\left(-\frac{E\alpha}{R}\right)$ used to compute the values of E_α and intercepts of these lines with the help of $g(\alpha)$ were used to find the values of frequency factor.

3.2 Thermodynamic Analysis

The apparent activation energy (E_α) obtained from the above DFM method and the degradation data collected by TGA studies were used to calculate the change in enthalpy (ΔH) [8]:

$$\Delta H = E_\alpha - RT_m \tag{5}$$

3.3 Reaction Mechanism: Master Plots

For a step-by-step procedure with an unchanging $g(\alpha)$ expression, a master plot based on the integral form of the kinetic data [9] analysis presents a perfect choice of the right kinetic model. From the integral form of the reaction model, $g(\alpha)$, using a reference point of $\alpha = 0.5$ and using the approximation of Senum and Yang [13], one can get the following equation:

$$\frac{g(\alpha)}{g(0.5)} = \frac{p(u)}{p(0.5)} \tag{6}$$

where $g(0.5) = \frac{k_0 E_\alpha}{\beta R} p(0.5)$ and $p(u) = \frac{e^{-u}}{u(u+4)(u^2+6u+6)}$, here $u = \frac{E_\alpha}{RT}$

The plot of $\frac{g(\alpha)}{g(0.5)}$ versus α against matches to theoretically plotted curves of various $g(\alpha)$ reactions. For the experimentally calculated plots of $\frac{p(u)}{p(0.5)}$ versus α, data from any heating rate can be used. The experimental master plot isn't affected by the heating rate. Equation (6) reveals that when a suitable kinetic model is applied, the experimentally calculated values of $\frac{p(u)}{p(0.5)}$ and the theoretically computed values of $\frac{g(\alpha)}{g(0.5)}$ are comparable. This method can be used to figure out how decomposition reactions work.

4 Results and Discussion

4.1 *Thermal Analysis*

Figure 1a shows the TGA results for polypropylene grocery bags, while Fig. 1b shows the DTG results. This image shows typical TGA curves for polypropylene grocery bags heated at various rates in a nitrogen atmosphere (see Fig. 1a). During the reaction with nitrogen, each weight loss curve is smooth and has two inflection points. All investigations used a final temperature of 900 °C, and the TGA curve patterns are almost identical at each of the four heating rates. In the temperature range of 30–900 °C, polypropylene degrades in two phases, as shown in the study in Fig. 1b. The DTG curve showed two peaks, which is noteworthy. The zone-1 due to polypropylene fusion, which occurs between 350 and 550 °C, and the zone-2 is due to the energy required to reorganize the structures formed during the macromolecule cracking process, which occurs at temperatures over 650 °C.

4.2 *Kinetic Analysis*

Figure 2 shows the kinetic curves of polypropylene grocery bags for DFM technique. The gradient of the kinetic plots was used to determine the E_α values. As shown in Table 1, the apparent activation energy was estimated based on the gradients of the isoconversional kinetic plots. In order for plastics to degrade, the C–C bonds that link the polymer chain must be disrupted. Weaker links in polymers were more easily thermally broken when E_α was lower during lower conversion, whereas the increasing tendency of activation energy with conversion tells about the energy needed to break the carbon–carbon bonds in polypropylene supermarket bags required at higher temperatures [14]. The average value of E_α obtained by using

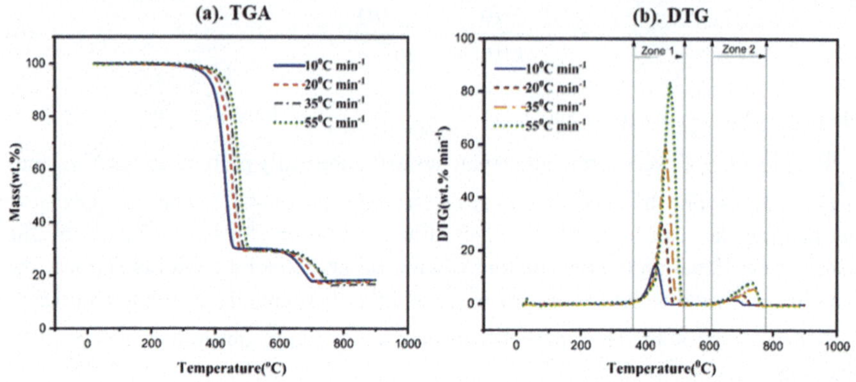

Fig. 1 TG and DTG plots form polypropylene grocery bags at four heating rates

DFM method for PP bags was 158.27 kJ mol^{-1}. The average value of E_α for PP bags obtained through DFM isoconversional procedure is comparable to data reported for other types of waste plastics. According to Al-Salem and Lettieri [15], E_α values for PET, PS, and LDPE were 181.02,159.15, and 159.15 kJ/mol, respectively. Das and Tiwari [16] discovered that E_α values of PP, LDPE, and HDPE range from 124.1 to 187.2 kJ/mol, 148.3 to 256.4, and 134 to 258, respectively. Singh et al. [17] investigated pyrolysis kinetics of waste milk packets, with E_α values ranging from 134.1 to 299.5 kJ/mol. It has been suggested that using differential methods can be quite incorrect and not very exact. As a result, the DFM approach was inaccurate because the thermogravimetric data is subjected to numerical differentiation, which introduces errors into the output. The average R^2 values derived by the DFM approach were 0.9843.

4.3 *Thermodynamic Analysis*

Equation 5 was used to determine the change in enthalpy (ΔH) for polypropylene grocery bags. The thermodynamic analysis was crucial to understand how ΔH changes at the level of conversion [18]. For all of the heating rates, one can use the values of E_α from the DFM to calculate ΔH. The average values of ΔH from the DFM at a heating rate of 10 K min^{-1} were 153.11 kJ/mol. In this study, there was no substantial change in the values of ΔH for when heating rates were changed. It should be noted that, given a particular conversion value, the difference between the average values of ΔH and E_α varied within a 5–6 kJ/mol range, as shown in Table 1. In accordance with this, it is worth highlighting that a slight difference between ΔH and E_α provides ideal circumstances for the creation of the activated complex [19]. The small variation between ΔH and E_α represents the simplicity of reaction, i.e., products may be formed with a small amount of energy [17, 20].

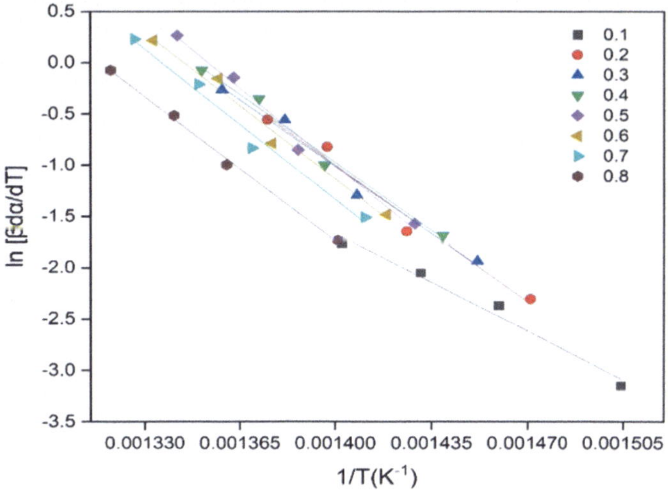

Fig. 2 Kinetic plot for polypropylene using DFM model

Table 1 Comparison of activation energy and enthalpy for different conversion values

Conversion	Activation energy (E_a)	Enthalpy (KJ/mol)
0.1	112.96	107.10
0.2	158.37	152.51
0.3	154.78	148.92
0.4	156.85	150.99
0.5	178.83	172.97
0.6	169.53	163.67
0.7	174.02	168.16
0.8	166.48	160.63

4.4 Prediction of Reaction Mechanism

By using Eq. (6), the $P(u)/P(0.5)$ values can be estimated by applying the pre-set apparent activation energy, and the activation energy values applied in this work were those acquired by the DFM approach and also utilizing the Senum and Yang 2nd approximation for $P(u)/P(0.5)$. The graphs of $g(\alpha)/g(0.5)$ were generated by combining many different kinetic models. Figure 3 depicts theoretically calculated master plots of various reaction mechanisms as well as experimentally calculated curves for PP bags for 10 K/min heating rate. $g(\alpha)/g(0.5)$ shows that polypropylene grocery bags degraded thermally under the influence of the zero-reaction model (F0) reaction mechanism at all heating rates, as can be seen from the experimental and theoretical master curves shown above. In light of the fact that the $P(u)/P(0.5)$ values for four different heating rates were nearly identical, these findings imply that

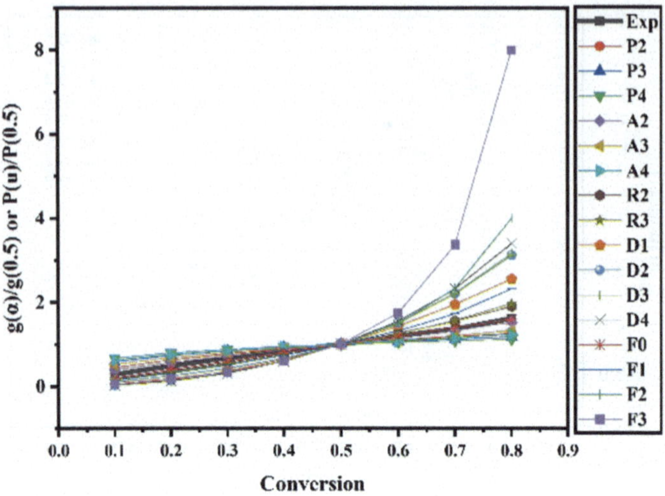

Fig. 3 Master plots based on the integral form of the kinetic data for DFM method at 10 K/min heating rate

the degrading response mechanism could be considered irrespective of heating rate employed [21].

5 Conclusions

The goal of this study was to use isoconversional methods and TGA data to figure out the kinetic parameters for pyrolysis of the polypropylene grocery bags. The E_α values were between 112 and 166 kJ/mol. The master plot made with the integral method made it clear that the F0 model can accurately describe how polypropylene grocery bags react. The slight difference between E_α and ΔH indicated how easily the reaction can occur provided the supplied energy is in the range of values of activation energy and enthalpy. Lastly, this analysis showed that the pyrolysis method can be used to treat polypropylene grocery bags, which will help reduce waste and recover the energy.

References

1. D.P. Serrano, J. Aguado, J.M. Escola, Catalytic cracking of a polyolefin mixture over different acid solid catalysts. Ind. Eng. Chem. Res. **39**(5), 1177–1184 (2000). https://doi.org/10.1021/ie9906363
2. I. Adam, T.R. Walker, J.C. Bezerra, A. Clayton, Policies to reduce single-use plastic marine pollution in West Africa. Mar. Policy **116**, 103928 (2020). https://doi.org/10.1016/J.MARPOL.2020.103928

3. S. Vyazovkin, D. Dollimore, *Linear and Nonlinear Procedures in Isoconversional Computations of the Activation Energy of Nonisothermal Reactions in Solids* (1996)
4. S. Singh, J. Prasad Chakraborty, M. Kumar Mondal, Intrinsic kinetics, thermodynamic parameters and reaction mechanism of non-isothermal degradation of torrefied Acacia nilotica using isoconversional methods. Fuel. **259** (2020). https://doi.org/10.1016/j.fuel.2019.116263
5. D. Rammohan, N. Kishore, R.V.S. Uppaluri, Insights on kinetic triplets and thermodynamic analysis of delonix regia biomass pyrolysis. Bioresour. Technol. 127375 (2022). https://doi.org/10.1016/J.BIORTECH.2022.127375
6. D. Rammohan, N. Kishore, R.V.S. Uppaluri, Reaction kinetics and thermodynamic analysis of non-isothermal co-pyrolysis of *Delonix regia* and tube waste. Bioresour. Technol. Reports **18**, 101032 (2022). https://doi.org/10.1016/J.BITEB.2022.101032
7. H.L. Friedman, Kinetics of thermal degradation of char-forming plastics from thermogravimetry. Application to a phenolic plastic. J. Polym. Sci. Part C Polym. Symp. **6**(1), 183–195 (2007). https://doi.org/10.1002/polc.5070060121
8. V. Dhyani, J. Kumar, T. Bhaskar, Thermal decomposition kinetics of sorghum straw via thermogravimetric analysis. Bioresour. Technol. **245**, 1122–1129 (2017). https://doi.org/10.1016/J.BIORTECH.2017.08.189
9. L.A. Pérez-Maqueda, J.M. Criado, P.E. Sánchez-Jiménez, Combined kinetic analysis of solid-state reactions: a powerful tool for the simultaneous determination of kinetic parameters and the kinetic model without previous assumptions on the reaction mechanism. J. Phys. Chem. A **110**(45), 12456–12462 (2006). https://doi.org/10.1021/jp064792g
10. Y. Chen, Q. Wang, Thermal oxidative degradation kinetics of flame-retarded polypropylene with intumescent flame-retardant master batches in situ prepared in twin-screw extruder. Polym. Degrad. Stab. **92**(2), 280–291 (2007). https://doi.org/10.1016/j.polymdegradstab.2006.11.004
11. P.K. Roy, P. Surekha, C. Rajagopa, V. Choudhary, Thermal degradation studies of LDPE containing cobalt stearate as pro-oxidant. Express Polym. Lett. **1**(4), 208–216 (2007). https://doi.org/10.3144/expresspolymlett.2007.32
12. L.T. Vlaev, V.G. Georgieva, S.D. Genieva, Products and kinetics of non-isothermal decomposition of vanadium(IV) oxide compounds. J. Therm. Anal. Calorim. **88**(3), 805–812 (2007). https://doi.org/10.1007/s10973-005-7149-y
13. G.I. Senum, R.T. Yang, Rational approximations of the integral of the Arrhenius function. J. Therm. Anal. **11**(3), 445–447 (1977). https://doi.org/10.1007/BF01903696
14. A. Aboulkas, K. El harfi, A. El Bouadili, Thermal degradation behaviors of polyethylene and polypropylene. Part I: pyrolysis kinetics and mechanisms. Energy Convers. Manag. **51**(7), 1363–1369 (2010). https://doi.org/10.1016/J.ENCONMAN.2009.12.017
15. S.M. Al-Salem, P. Lettieri, Kinetic study of high density polyethylene (HDPE) pyrolysis. Chem. Eng. Res. Des. **88**(12), 1599–1606 (2010). https://doi.org/10.1016/J.CHERD.2010.03.012
16. P. Das, P. Tiwari, Thermal degradation kinetics of plastics and model selection. Thermochim. Acta **654**, 191–202 (2017). https://doi.org/10.1016/J.TCA.2017.06.001
17. G. Singh, A.K. Varma, S. Almas, A. Jana, P. Mondal, J. Seay, Pyrolysis kinetic study of waste milk packets using thermogravimetric analysis and product characterization. J. Mater. Cycles Waste Manag. **21**(6), 1350–1360 (2019). https://doi.org/10.1007/s10163-019-00891-9
18. Y. Xu, B. Chen, Investigation of thermodynamic parameters in the pyrolysis conversion of biomass and manure to biochars using thermogravimetric analysis. Bioresour. Technol. **146**, 485–493 (2013). https://doi.org/10.1016/J.BIORTECH.2013.07.086
19. L. Vlaev, I. Markovska, L. Lyubchev, Non-isothermal kinetics of pyrolysis of rice husk. Thermochim. Acta **406**(1–2), 1–7 (2003). https://doi.org/10.1016/S0040-6031(03)00222-3
20. A.K. Vuppaladadiyam, N. Merayo, P. Prinsen, R. Luque, A. Blanco, M. Zhao, A review on greywater reuse: quality, risks, barriers and global scenarios. Rev. Environ. Sci. Bio/Technology **18**(1), 77–99 (2019). https://doi.org/10.1007/s11157-018-9487-9
21. S.S. Ghadikolaei, A. Omrani, M. Ehsani, Non-isothermal degradation kinetics of ethylene-vinyl acetate copolymer nanocomposite reinforced with modified bacterial cellulose nanofibers using advanced isoconversional and master plot analyses. Thermochim. Acta **655**, 87–93 (2017). https://doi.org/10.1016/J.TCA.2017.06.014

GPSR Compliance

The European Union's (EU) General Product Safety Regulation (GPSR) is a set of rules that requires consumer products to be safe and our obligations to ensure this.

If you have any concerns about our products, you can contact us on ProductSafety@springernature.com

In case Publisher is established outside the EU, the EU authorized representative is:

Springer Nature Customer Service Center GmbH
Europaplatz 3
69115 Heidelberg, Germany

The manufacturer's authorised representative in the EU is Springer Nature Customer Service Centre GmbH, Europaplatz 3, 69115 Heidelberg, Germany. If you have any concerns regarding our products, please contact ProductSafety@springernature.com

Printed and bound by CPI Group (UK) Ltd, Croydon, CR0 4YY
01/10/2026
02225911-0002